The Paradox of Water

The publisher and the University of California Press Foundation gratefully acknowledge the generous support of the Ralph and Shirley Shapiro Endowment Fund in Environmental Studies.

The Paradox of Water

THE SCIENCE AND POLICY OF SAFE DRINKING WATER

Bhawani Venkataraman

UNIVERSITY OF CALIFORNIA PRESS

University of California Press
Oakland, California

© 2023 by Bhawani Venkataraman

Library of Congress Cataloging-in-Publication Data

Names: Venkataraman, Bhawani, 1964- author.
Title: The paradox of water : the science and policy of safe drinking
    water / Bhawani Venkataraman.
Description: Oakland, California : University of California Press, [2023] |
    Includes bibliographical references and index.
Identifiers: LCCN 2022019423 (print) | LCCN 2022019424 (ebook) |
    ISBN 9780520343436 (cloth) | ISBN 9780520343443 (paperback) |
    ISBN 9780520974791 (ebook)
Subjects: LCSH: Drinking water. | Drinking water--Safety measures.
Classification: LCC TD430 .V45 2023 (print) | LCC TD430 (ebook) |
    DDC 628.1028/9--dc23/eng/20220908
LC record available at https://lccn.loc.gov/2022019423
LC ebook record available at https://lccn.loc.gov/2022019424

32 31 30 29 28 27 26 25 24 23
10 9 8 7 6 5 4 3 2 1

For Jay

Contents

List of Illustrations	ix
Preface	xiii

1.	Introduction	1
2.	Liquid Water: An Essential Ingredient for Life	6
3.	Water: A Potential Threat to Life	19
4.	Why Drinking Water Quality Matters	30
5.	Making Water Safe	45
6.	Learning from Drinking Water Contamination Events	70
7.	The Precautionary Principle and Safe Drinking Water	88
8.	Protecting Nature: Ecosystem Services for Drinking Water	96
9.	Recycled Potable Water	110
10.	Decentralized, Appropriate Drinking Water Treatments	132
11.	Valuing Safe Drinking Water	154

Acknowledgments	167
Notes	169
Additional Resources	205
Index	209

Illustrations

FIGURES

1. Different representations of methane, ammonia, and water — 10
2. Representations of the electron distribution in methane, ammonia, and water molecules — 12
3. Attractive interactions between two H_2O molecules between regions of higher electron density in one molecule and lower electron density in another molecule — 13
4. Hydrogen bonding between water molecules forms a network of interactions — 14
5. Self-assembly of phospholipid molecules to form a bilayer that defines the cell membrane — 18
6. Data from the WHO showing global trends in reported cholera cases from 1985 to 2016 — 22
7. Data from the CDC showing trends in reported drinking water contaminations in municipal systems in the United States from 1971 to 2014 — 24

8.	Cartoons that appeared in the 1850s in *Punch*, a British magazine, highlighted public sentiment about the quality of the water in the Thames River	33
9.	A cartoon that appeared in a 1919 issue of *The American City* championing chlorine disinfection for treating drinking water	37
10.	Percentage of people across geographic regions with access to the water source and quality categories defined by the JMP drinking water ladder	41
11.	Fractions of water on Earth that make up oceans and fresh water	46
12.	Schematic of the steps involved in a drinking water treatment facility, using water from surface or groundwater sources	56
13.	Schematic of the processes and sectors of society that play a role in ensuring the delivery of safe and reliable drinking water to a community	59
14.	Percentage of PWS in the United States with acute health-based and health-based violations from 2011 to 2020	61
15.	Violations of federally mandated regulations by public water systems in the United States between 1982 and 2015	64
16.	A watershed comprises the land through which precipitation drains into local bodies of water, including lakes, rivers, oceans, and groundwater	99
17.	The hydrologic cycle, which dictates the movement of water across the Earth	111
18.	Schematic of "de facto" water reuse.	115
19.	Schematic of indirect potable reuse (IPR)	117
20.	Schematic of direct potable reuse (DPR)	118
21.	Pore sizes of different filtration steps to address a range of biological and chemical contaminants	122
22.	Schematic of the multibarrier approach	123
23.	An example of how sequential treatment steps in the multibarrier approach address chemical and biological contaminants so that the water that leaves the facility is high quality	124
24.	Impacts on the over 2 billion people (about 25% of the global population) lacking safe drinking water	133

25. The percentage of people in four locations exposed to *E. coli* contamination from water collected from a source and after this water is stored in homes — 135
26. Construction of a biosand filter — 141
27. Schematic showing the different layers of the Arsenic SONO filter — 145
28. Data comparing average per capita household water use across different nations. — 160

TABLES

1. Atmospheric composition and planetary conditions (atmospheric pressure and average surface temperature) of Venus, Earth, and Mars — 8
2. Physical properties of methane, ammonia, and water — 9
3. Examples of pathogens by type and disease transmitted through water — 21
4. The JMP "drinking water ladder" classifications, based on access and water quality — 40
5. Categories of contaminants regulated under the NPDWR and their sources, examples, and health effects — 51
6. Acronyms and names of fluorinated compounds — 83
7. Comparisons of the water management services provided by natural versus built infrastructure — 102
8. Examples of green infrastructure — 107
9. Examples of chemicals that may be present in wastewater and their potential sources — 113
10. Examples of pathogens that may be present in wastewater and the diseases they can cause — 114
11. Examples of indirect and direct potable reuse systems across the world — 119
12. Comparing the effectiveness and costs of the HWTS methods described in the text — 147

Preface

For over 15 years, I have been teaching at Eugene Lang College of Liberal Arts, The New School, in New York City. At The New School, we encourage students to explore contemporary, socially relevant issues through interdisciplinary lenses and to use their education to strive for a more socially just, equitable future for all. So, the questions I grappled with when I first came to The New School were: How can the teaching and learning of chemistry be achieved through socially relevant contexts? What topics would engage students? What current issues rely on chemical perspectives to inform solutions and policies that demonstrate the relevance of chemistry? But, at the same time, the topic should clearly illustrate the importance of drawing from multiple disciplines in informing approaches and policies, and issues of justice and equity must be central to the topic.

I also drew from the chemical education research literature that indicates that students may not recognize why chemistry matters in their everyday lives. Chemistry can be challenging to learn as it deals with a scale beyond the human senses. We often learn by seeing and interacting. However, this is not possible with molecules. To a chemist, the significance of the molecular scale may seem evident as everything around us is a result of molecular-scale interactions, from complex cellular processes to the plastics used in water bottles. However, to students, this may not be immediately obvious.

As I began developing an introductory undergraduate chemistry course, I researched contexts where chemistry is central, requires interdisciplinary connections, and is socially relevant. Through my teaching, I have realized that the global challenge around access to safe drinking water is one such context. We all have a relationship with water. We know we need water—we drink it, cook with it, and bathe in it. Water evokes cultural and religious sentiments. However, from a public health perspective, it is not just water that humans need, but safe drinking water. Access to safe drinking water allows basic needs to be met, supports educational opportunities for children, helps overcome gender inequities, lowers the stress and anxiety of families, and allows for more socially and economically productive uses of time. As a result, one of the United Nations' Sustainable Development Goals focuses on access to safe water and sanitation for all.[1]

As this book aims to demonstrate, it is the chemistry of water that makes it both essential for life and at the same time easily susceptible to contamination. At the molecular level, water displays complex properties. These properties are dictated by the hydrogen and oxygen atoms that form the molecule described as H_2O. This deceptively simple formula, H_2O, dictates the properties that make water essential for life on Earth. Students know this but often do not understand why. So, learning and appreciating the chemistry of water is a way for students to see how the molecular scale is relevant to their very existence. Ironically, water's chemical properties that make it essential for life are also why water is so easily contaminated and potentially a threat to life.

This paradox—water being essential for life but easily contaminated and hence a potential threat to life—is in part a consequence of its chemistry. Recognizing these chemical properties of water brings about a more nuanced understanding of the challenges around access to safe drinking water. In concert with biological and physical as well as social, economic, and political factors, this chemical understanding is crucial to informing drinking water treatments and the regulatory frameworks relevant to the delivery of safe drinking water.

Most people living in the Global North take access to safe drinking water for granted, but this is not necessarily the case in the Global South. Why is this? Exploring these questions requires understanding the chemistry of water *and* historical, economic, social, cultural, and political factors, such as the impacts of colonialism. Even in the Global North, while the majority have access to safe drinking water, it is certainly not universal; understanding why is related to the chemistry of water as well as the inequities that dominate society. For example, the fact that the drinking water delivered to residents in Flint, Michigan, had unsafe levels of lead can be understood at the chemical level—what happened to make lead leach from the pipes. But why the residents of Flint were exposed to unsafe levels of lead in their drinking water is a consequence of systemic racism.[2] Flint is just one of many cities dealing with unsafe drinking water in the United States, a country that prides itself on being "developed." Data reveal that those most impacted by unsafe water are marginalized, low-socioeconomic communities.[3,4] The chemistry of water makes the delivery of safe drinking water complex and expensive, while also requiring significant expertise. As a result, safe water is often too expensive for many communities. The COVID-19 pandemic has brought into sharp focus the importance of safe water to protect public health.

I was working on this book in spring 2020 as the world was becoming aware of the SARS-CoV-2 virus, that is, the coronavirus. As the gravity of this pandemic unfolded, we were constantly reminded to wash our hands, scrubbing thoroughly with soap for 20 seconds (this amounts to approximately two liters, or roughly half a gallon, of water if you keep the tap running for 20 seconds). We were even advised to rinse out groceries as soon as we brought them into our homes and consider wiping down mail and newspapers. For most people living in the Global North, the one thing that we did not have to be concerned about was the safety of the water coming out of our taps.

Reading these handwashing instructions broadcast to the public, I could not help but wonder about the challenges faced by communities that do not have access to safe water. If a way to protect yourself from the virus is to wash your hands with soap and water, what do you do if you do not have safe water or running water at home? A *New York Times* opinion piece poignantly described a woman living on the Navajo reservation in

Arizona who did not have indoor plumbing.[5] Her son traveled about 90 minutes away to collect water to bring to his mother. Unfortunately, it appears that the son contracted the virus during such a trip. Both he and his mother died from COVID-19. Access to piped, safe water might have saved their lives. Globally, similar stories are repeating in communities that lack access to safe water. This pandemic has affected almost every person on Earth, although historically marginalized communities have borne the brunt once again. The COVID-19 pandemic has put into sharp focus the social and economic consequences when public health is at risk. Even before this pandemic, millions of people worldwide have been dealing with health risks from unsafe water and its vast toll on communities' health, social, economic, and educational outcomes.

In my experience as an educator, helping students understand the chemical principles that make water both essential to life and potentially a threat creates respect for the role of water in our existence and the crucial role of safe drinking water to our well-being. This book aims to help readers understand why the chemistry of water placed within social, political, cultural, and economic perspectives must inform policies, solutions, and actions that ensure sustainable and equitable access to safe drinking water for all.

1   Introduction

If you live in a country in the Global North,[1] you most likely open the tap, fill your glass, and drink the water. You probably do this reflexively, not pausing to ask, "Wait, will drinking this water or cooking with it make me or my family sick?" Being able to do so without hesitation is what it means to have access to safe drinking water—a fundamental human right according to the United Nations. While it is true that water is essential for life, it is safe drinking water that is necessary to protect public health and support social and economic development. As defined by the World Health Organization (WHO), safe drinking water "does not represent any significant risk to health over a lifetime of consumption, including different sensitivities that may occur between life stages."[2] Access to safe drinking water means you should not have to worry about getting sick or dying from your tap water. Not having to worry about the safety of the water you drink is worth valuing and championing.

But, the next time you drink a glass of tap water, pause and ask yourself, "Why can I drink this water and not worry about its safety?" You may have heard that other regions of the world cannot make this assumption. Over two billion people worldwide (about one in four people) consume water that can potentially cause them to become sick; over 800,000 people die from waterborne diseases every year. Then ask yourself the following: Have I recently heard of a town in the United States where the tap water was deemed unsafe? Am I aware of the following incidents of contamination of drinking water in communities in the United States?

> A pediatrician in Flint, Michigan, informs a mother that her child's blood lead level has increased since the last test. And the pediatrician suspects that the tap water delivered to this family's home is the likely cause.[3]
>
> A mother in California, after hearing about the community in Flint exposed to unsafe levels of lead from the tap water, decides to look at her town's drinking water quality report only to read the fine print that said "1,2,3-Trichloropropane has been detected in 29 wells in Fresno. Some people who use water containing it over many years may have an increased risk of getting cancer, based on studies in laboratory animals."[4]
>
> A man in Hoosick Falls, New York, wonders if his father's death from cancer may have been due to the presence of a chemical called PFOA that was used in local industries and detected in the town's drinking water.[5]
>
> Harmful algal blooms in freshwater sources have released toxic chemicals called cyanotoxins, contaminating drinking water sources and posing a health threat to people and aquatic ecosystems.[6,7]
>
> The toxic solvent trichloroethylene, used by an aerospace industry in Long Island, New York, has been leaching into the groundwater, the drinking water source for local communities. Residents wonder if the cancer cases in their community may have been caused by drinking their tap water for years.[8]
>
> In Appalachia, communities where coal mining supports the local economy consume drinking water that contains unsafe levels of toxic metals and other chemical contaminants. These chemical contaminants result from runoff from the mines seeping into drinking water sources.[9]
>
> In some communities in the San Joaquin Valley, California, the tap water is contaminated by nitrate and pesticides used to grow the produce that feeds the country.[10,11] Many of these communities are low-socioeconomic and marginalized and have to buy bottled water, consuming a significant fraction of their incomes.
>
> Over two million people in the United States, disproportionately Native American and low-socioeconomic communities, have been dealing for decades with the impact of unsafe water on their health. Many in these communities lack piped water and travel a long distance to collect and bring home water.[12,13,14]

As forest fires spread into towns, toxic chemicals are released into the air and detected in local drinking water sources.[15]

And the list goes on . . .[16,17,18]

Why did the water in these towns get contaminated? Are these isolated events unlikely to be repeated, or are such incidents increasing? Why is it that some communities in the United States have been dealing with unsafe water for years? Why do over two billion people across the globe consume unsafe water? As explored in this book, the reasons are manyfold, but at heart is the chemistry of water, which is the key to identifying and understanding water contamination.

Water is a molecular marvel. Its seemingly simple molecular formula—H_2O—contradicts its complex behavior. The fact that water molecules are made up of two hydrogen atoms and one oxygen atom, that is, H_2O, allows life to thrive under the conditions that exist on Earth. Every living organism is connected through this reliance on liquid water, from an amoeba to a plant to a blue whale. It is, however, also because water is H_2O that water is easily contaminated. We cannot live without water, and at the same time, water being easily contaminated can therefore potentially threaten life—this is the paradox of water. Intentionally acknowledging this paradox of water is crucial to addressing the many challenges faced in ensuring access to safe drinking water worldwide. In nations where the majority have access to safe drinking water, this very success has made us complacent. This complacency is dangerous and threatens to unravel this success, as is evident in the increasing incidences of water contamination reported in the United States.

Due to the inherent chemistry of water, the quality of water from precipitation to reservoir to tap changes. A drinking water source may originate in pristine landscapes, mountain springs, or snowmelts. This water may flow through farmland and industrial zones and percolate into the ground before emptying into a reservoir. That this water, which has likely been subject to pollution from rampant industrialization and agriculture, can be rendered safe enough to drink at the tap demands constant investment and oversight. Understanding the chemistry that defines first the processes of water contamination and then treatment and quality control necessary for the delivery of safe drinking water allows for informed water management decisions and a vigilant public.

In the early 20th century, our understanding of how water gets contaminated led to scientific, engineering, health, economic, and policy investments in delivering safe water. As a result, people living in nations that could afford to make the necessary investments benefited enormously. The number of people getting sick or dying from waterborne diseases like cholera and typhoid plummeted. Economic productivity grew. But this, unfortunately, is still not the case for a significant fraction of the world's population. This book aims to awaken the sensitivity of 21st-century readers to a deeper understanding of the most fundamental aspects of water, arguing that the chemistry of water demands urgent responses in increasing investments for water infrastructure and strengthening regulations to ensure the continued delivery of safe water. Ignoring this chemistry could unravel the gains enjoyed by the majority and make it even more unlikely to address the challenges faced by the minority unjustly left out from enjoying the benefits of safe water.

This book begins by exploring the fundamental chemistry of water that dictates the paradox of water. Chapter 2 examines why water's molecular formula (H_2O) dictates its unique properties and makes water essential for life. Chapter 3 explores why the same properties of water that make it essential for life also allow water to be easily contaminated and potentially life-threatening. This introduction to the chemistry of water emphasizes why understanding water's molecular nature is crucial in guiding the social and policy dimensions of access to safe drinking water.

The history of human civilization is intricately connected to access to water sources. As populations rose, so did pollution of water sources, resulting in the rampant spread of diseases and high mortality and morbidity rates. It took a scientific understanding of why water was responsible for spreading diseases to develop informed solutions to treat water, resulting in enormous benefits to communities. As a result, mortality rates declined, and economic productivity increased. Chapter 4 traces this historical development to explore the intersection of access to safe drinking water with public health and the social and economic impacts of unsafe water.

Using the United States as a case study, chapter 5 explains the invisible path that water takes from source to tap and discusses the role of the Environmental Protection Agency (EPA) and congressional acts in regulating drinking water quality. These acts dictated investments in drinking water infrastructure, allowing the vast majority of people in the United States to have safe drinking water coming out of their taps. Finally, we will look at the successes and challenges facing the United States' regulatory process and growing risks that demand urgent attention to ensure the continued delivery of safe drinking water.

Chapter 6 traces case studies of water contamination and highlights the tragic consequences of disregarding water's chemical behavior, which have led to social and economic costs on communities. We must learn from these cases to inform how we strengthen drinking water management. The chemistry of water dictates that the precautionary principle must inform the management and delivery of safe drinking water. Chapter 7 explores this principle and how it applies to forward-thinking and climate-resilient solutions for safe drinking water management, as discussed in chapters 8 and 9. Examples of decentralized, appropriate solutions that respect local social, cultural, and economic realities in delivering safe water are presented in chapter 10.

Finally, chapter 11 highlights the importance of an informed public to ensure that their governments invest in the necessary infrastructure for safe drinking water. The more we understand the significance of safe drinking water to our health and well-being, the more informed our demands for sound and just actions by our political representatives will be.

No single book can do justice to the multitude and complex ways that water and society are intertwined. As a result, this book cannot cover all aspects of water and focuses on drinking water. Drinking water is the smallest fraction of water used by a nation; agricultural and industrial uses of water are higher. Yet, the quality of this smallest fraction is what dictates whether a community is healthy, educated, and economically sustained. Further, how we use or misuse water has consequences for all life on Earth. We must heed and respect the Lakota phrase *Mní wičhóni*, water is life.

2 Liquid Water

AN ESSENTIAL INGREDIENT FOR LIFE

All the rocky planets in the solar system—Mercury, Venus, Earth, Mars—had the same origin roughly 4.6 billion years ago. The gradual accretion of dust particles allowed these planets to grow, aided by the bombardment of meteors and other celestial objects. Yet today, Earth is different from the others. In no small part, life transformed Earth.

As far as we know, Earth is the only body in our solar system that harbors any form of life. So what is it about Earth's environment that allows it to support life and of such complexity and diversity? Scientists believe that the essential ingredients for establishing life (at least life as we know it) are liquid water, energy, and a source of carbon. Energy in the form of solar radiation is abundant on the rocky planets, as are sources of carbon. For example, a recent analysis of soil samples on Mars revealed the presence of carbon-containing compounds.[1] But only one of these planets supports liquid water.

The temperature of a planet dictates whether water is present in the liquid phase. The distance from the Sun and the planet's atmospheric composition determine its temperature. Greenhouse gases (GHGs) such as carbon dioxide, methane, and water vapor influence a planet's temperature through the greenhouse effect.[2] A planet absorbs solar radiation from the Sun. The planet warms up and radiates energy back into its atmosphere. GHGs in the atmosphere absorb some of this radiation from the planet. These GHGs then release this energy into the atmosphere, some of which is reabsorbed by the planet, warming it further. This greenhouse effect raises the temperature of a planet higher than it would otherwise be.

Comparing Earth's atmosphere to neighboring planets, we find significant differences (see table 1). The very high temperature on Venus is due to its proximity to the Sun and amplified by its atmosphere's high carbon dioxide levels, a greenhouse gas. The much colder temperature on Mars is related to its distance from the Sun and lower carbon dioxide levels in the Martian atmosphere (the atmosphere of Mars is 95% carbon dioxide, but its atmospheric pressure is 1/100 that of Earth's). While the average surface temperature of Mars is too low to support liquid water, recent studies suggest the possibility of liquid water under polar ice sheets.[3,4] However, scientists speculate that if water did exist under these ice sheets, the temperature of this water might be about −68°C, possibly due to dissolved salts, which lower the freezing point of water. Even if life existed in this Martian water at such frigid temperatures, it is hard to imagine this life would display the complexity and diversity of life on Earth.

On the other hand, Earth's atmosphere is "just right." The atmospheric levels of greenhouse gases like carbon dioxide and methane influence the average surface temperature of Earth. If these gases were not present, Earth's average surface temperature would be a frigid −18°C (around 0°F). At this temperature, water will freeze, and life might not have evolved on Earth.

It is not just that water is present on Earth but that the water is in a *liquid* state. Geological evidence suggests that liquid water may have formed on Earth as far back as 4.3 billion years ago.[5] So, 4.3 billion years ago, the essential ingredients of life—energy, sources of carbon, and liquid water—were present on Earth, ultimately leading to the chemistry that resulted in the planet's first living organism. It is unclear when and how the first forms of life evolved on Earth. Rock samples dated 3.46 billion years reveal the presence of ancient microfossils.[6] Some studies suggest that life's key biomolecules may have formed as early as 4.1 billion years ago.[7] The point is that once conditions on Earth supported liquid water, life took a foothold.

Table 1   Atmospheric composition and planetary conditions (atmospheric pressure and average surface temperature) of Venus, Earth, and Mars

	Venus	Earth	Mars
Atmospheric composition			
Nitrogen	4%	78%	2.7%
Oxygen	trace	21%	0.13%
Argon	0.01%	0.9%	1.6%
Carbon Dioxide	96%	0.04%	95%
Methane	0%	0.00019%	0%
Planetary conditions			
Atmospheric pressure relative to Earth	90	1	1/100
Average surface temperature	462°C	15°C	−63°C

WHY DOES EARTH SUPPORT LIQUID WATER?

Ever since scientists became aware of a liquid on Saturn's moon, Titan, comparisons have been made with Earth.[8] As stated in an article in the *New York Times*, "Like Earth, it [Titan] has a thick atmosphere of mostly nitrogen (the only moon that has much of an atmosphere at all), and like Earth, it has weather, rain, rivers and seas."[9] While these are similarities between Earth and Titan, the chemical composition of the rain, rivers, and seas are different. On Earth, this liquid is water; on Titan, this liquid is methane. The average surface temperature of Earth is 15°C (around 59°F), which supports liquid water; at this temperature, methane is a gas. The average surface temperature of Titan is a frigid −179°C (−290°F). Water is solid at this temperature, but methane is a liquid, hence the methane rain on Titan. While Titan may have liquid on its surface, liquid methane is not liquid water. While one cannot rule out that Titan harbors life, any life on Titan is unlikely to display the level of complexity and diversity of life on Earth. What is it about water, specifically liquid water, that allows life to thrive from the smallest of single-cell organisms to the

Table 2 Physical properties of methane, ammonia, and water

	Methane (CH_4)	Ammonia (NH_3)	Water (H_2O)
Boiling point (°C)	−162	−33	100
Melting point (°C)	−182	−78	0
Relative mass (amu)	16	17	18

massive dinosaurs, to us humans? To answer this question, we need to explore the molecular nature of water and methane.

Chemists compare molecules of similar size and mass as a way of understanding how the atoms that make up a molecule influence its physical and chemical behavior. Let's compare water, methane, and ammonia as these molecules' molecular size and mass are comparable. Table 2 lists the boiling and melting points and relative masses of these three molecules. As evident in table 2, the three molecules have very different boiling and melting points.

Water, ammonia, and methane all have hydrogen atoms, but water has an oxygen (O) atom; ammonia, a nitrogen (N) atom; and methane, a carbon (C) atom. The O, C, and N atoms do not differ in their atomic structure by much—the carbon atom has one less proton than nitrogen, which has one less proton than oxygen. As a result, the carbon atom has one less electron than nitrogen, which has one less electron than oxygen. These seemingly minor differences determine how many hydrogen atoms bond to carbon, nitrogen, and oxygen, dictating the chemical bonding between the atoms that make up these molecules. Figure 1 shows ways of representing these three molecules: the name in the top row, the molecular formula in the middle row, and the three-dimensional (3D) position of the atoms that make up the molecule in the bottom row. As the molecular formulas indicate, in methane, CH_4, the carbon atom is chemically bonded to four H atoms; in ammonia, NH_3, the nitrogen atom is chemically bonded to three H atoms; in water, H_2O, the oxygen atom is chemically bonded to two H atoms.

A chemical bond is a result of interactions between the electrons and protons of atoms. For example, in H_2O, the attraction between the

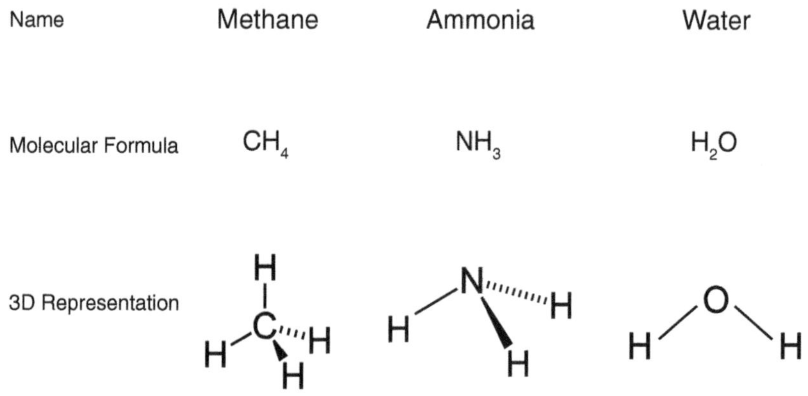

Name	Methane	Ammonia	Water
Molecular Formula	CH_4	NH_3	H_2O
3D Representation			

Figure 1. Different representations of methane (CH_4), ammonia (NH_3), and water (H_2O). The 3D representations show the relative positions of the atoms that make up a molecule. In these 3D representations, the dashed line (between C and an H and N and an H) indicates that this hydrogen atom is behind the plane of the page; the wedge indicates the hydrogen atom is in front of the plane of the page.
Illustration by Ryann Abunuwara

oppositely charged electrons and protons of the two Hs and the one O form a chemical unit. The chemical bonds between the H and O atoms are represented as H-O-H. Similarly, in NH_3, each of the three Hs forms a chemical bond with N, and in CH_4, each of the four Hs forms a bond to the C atom. This chemical bonding dictates the molecular shapes, or geometries, shown in figure 1, as indicated by the 3D representations. As suggested by the 3D representations, methane has a "tetrahedral" geometry, ammonia a "trigonal pyramidal" geometry, and water a "bent" geometry. If you could imagine drawing a line that connects the nuclei of the H-O-H atoms, this angle would be less than 180°. The nature of the atoms that make up these molecules dictates the chemical bonding between atoms and, as a result, the differences in their molecular shape. These distinctions between the three molecules result in the vastly different boiling and melting points listed in table 2.

The fact that water contains an oxygen atom is also significant. All atoms exhibit an atomic property called electronegativity. This property influences the electron distribution in molecules. In the case of the water molecule, the O atom has a high electronegativity, in fact, the second-highest of all atoms. As a result, the electrons that form the chemical bond between each H and the O experience a stronger attraction from the O atom than the H atom. Consequentially, the H_2O molecule's electron distribution is skewed such that the O end of the H_2O molecule has a higher electron density than the two H ends. Figure 2 depicts the electron distributions in the three molecules CH_4, NH_3, and H_2O. As can be seen in the depiction of the electron density for H_2O, the darker area indicates higher electron density on the O atom in H_2O. The lighter areas indicate lower electron density over the two H atoms. This asymmetric electron distribution and its bent structure make H_2O a *polar* molecule. The N atom is electronegative but less so than the O atom. This lower electronegativity and the trigonal pyramidal structure of NH_3 make ammonia less polar than water. In figure 2, this is depicted by the relatively lighter shading (less "black") around the N atom in NH_3 compared to that for the O atom in H_2O, and the hydrogen atoms in NH_3 appear less "white" compared to the H atoms in H_2O. The C atom is even less electronegative than the N atom, which, along with the symmetrical molecular structure of CH_4, makes the methane molecule *nonpolar*. In other words, the electron distribution in the CH_4 molecule is distributed symmetrically around all the atoms.

A molecule's electron distribution and its three-dimensional molecular structure dictate its polarity. Water is polar; ammonia is also polar but less than water; methane is nonpolar. As a result of this polarity, two water molecules experience attractive interactions (see figure 3). The attractive interaction between polar molecules is similar to magnets, where opposite poles of two magnets experience attractive interactions. Two ammonia molecules also experience attractive interactions. But, since ammonia is less polar than water, the attraction between ammonia molecules is weaker. And just as it takes more energy to separate two stronger magnets than two weaker magnets, it requires more energy to separate two water molecules than two ammonia molecules. Nonpolar molecules like CH_4 also experience interactions between molecules, but these interactions are weaker than between polar molecules like water and ammonia.

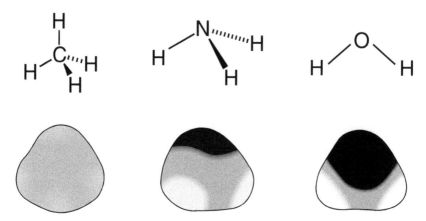

Figure 2. Representations of the electron distribution in methane (left), ammonia (middle), and water (right) molecules. Water and ammonia are polar, as indicated by the dark regions of higher electron density (O and N, respectively) and light regions of lower electron density (the H atoms). Methane is nonpolar, as indicated by the symmetric coloration of the electron density. Water's higher polarity compared to ammonia's is indicated by the darker regions around the O compared to the N and the lighter regions of the H atoms in H_2O compared to the H atoms in NH_3. (For ammonia, in the view represented here, two of the three H atoms are visible; the third is behind the plane of the page.)

Illustration by Ryann Abunuwara

Now imagine a cup of water and a cup of ammonia. While the molecules in each cup are constantly moving around, water molecules are attracted to each other; the same for ammonia molecules. The more polar the molecules, the stronger the interaction between molecules. As a result, the collection of water molecules will, on average, experience stronger interactions between each other as compared to the ammonia molecules. Now say you want to boil this cup of water, so you heat it on a stove. To determine when the water boils, you wait to observe bubbles in the water. The bubbles are the gaseous water molecules that have absorbed enough energy to overcome the attraction between water molecules and move into the vapor phase.

The temperature at which a liquid boils is related to the energy required for molecules to overcome the attraction between other molecules and

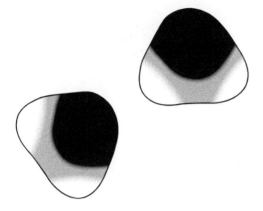

Figure 3. Attractive interactions between two H_2O molecules between regions of higher electron density in one molecule (around the O atom, darker shading) and lower electron density in another molecule (around an H atom, lighter shading).

Illustration by Ryann Abunuwara

escape into the gas state. The stronger the interactions between molecules, the higher the boiling point. The more polar the molecule (comparing molecules of similar size), the more energy is required to overcome the attraction between molecules, and the higher the boiling point. Consequentially, water has the highest boiling point of the three molecules in table 2. Ammonia is less polar than water and has a lower boiling point.

Water is also distinctive in that water molecules interact with each other through a particular type of polar interaction called hydrogen bonding. Hydrogen bonding is a consequence of polarity and other characteristics of the atoms that make up the water molecule. A result of hydrogen bonding is that water molecules form an intricate "network" of interactions between neighboring molecules (see figure 4). To move from the liquid to the gas phase, a water molecule must absorb enough energy to overcome this network of attractive interactions between neighboring molecules. Ammonia molecules also form hydrogen bonds between each other. However, ammonia's lower polarity and its molecular structure and shape do not allow for the intricate, strong network of interactions formed between water molecules. The consequence of these seemingly abstract concepts is very real. Water boils at 100°C; ammonia boils at –33°C. Therefore, water is liquid on Earth's surface, but ammonia is a gas. Being nonpolar, methane exhibits weak interactions between molecules, resulting in a low boiling point (–161°C), and is a gas at conditions on Earth's surface.

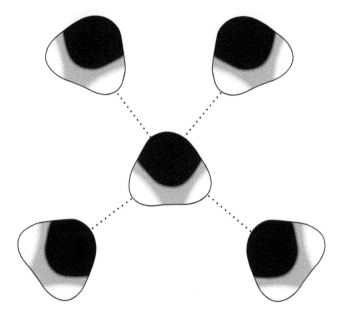

Figure 4. Hydrogen bonding (denoted by dotted lines) between water molecules forms a network of interactions. The hydrogen bonding interaction is between the O in one H_2O molecule and an H in a neighboring H_2O molecule. Each of the outer H_2O molecules shown in the figure interacts with neighboring H_2O molecules (not shown in the figure).

Illustration by Ryann Abunuwara

SO, IS IT JUST THE BOILING POINT OF WATER THAT MATTERS FOR LIFE?

In addition to its high boiling point, water also has a relatively high melting point because it is a bent, polar molecule that hydrogen bonds. The polar water molecules in ice—solid phase—interact with neighboring water molecules. As with water molecules in the liquid phase, in the solid phase, the water molecules are, on average, oriented to maximize attractive interactions. For ice to melt, energy must be absorbed to separate the water molecules from each other to overcome the strong hydrogen-bonding networked interactions between water molecules. As a result, water melts at a relatively high temperature. Other than regions where temperatures are below 0°C, water on Earth's surface is liquid.

Another distinctive property of water is its high specific heat capacity. It takes a lot of energy to change the temperature of water (this is why water is used as a coolant). Since 70 percent of Earth's surface is covered by oceans, water's high heat capacity helps moderate climate (also why coastal cities have more moderate temperature changes). Simultaneously, water can exist in all three phases on Earth, which is crucial for the hydrologic cycle, allowing for evaporation and precipitation and the transportation of water across the globe. By comparison, ammonia will be gas in most regions of Earth's surface (except perhaps during the winters in the Arctic and Antarctica, where it may be a liquid), and methane will not be a liquid on Earth's surface. Water also exhibits high cohesive and adhesive forces, essential for biological functioning. For example, trees take advantage of this property to help transport water to leaves.

Another anomalous property is that water's solid form, ice, is less dense than liquid water. As we all know, ice floats in a glass of liquid water. In the solid phase, as in the liquid phase, water molecules interact with neighboring water molecules through the same polar, hydrogen-bonding interactions. However, in the solid phase, it turns out that optimizing these interactions results in a more open structure for the organization of water molecules compared to the organization of water molecules in the liquid phase. As a result, when comparing the same volume of solid water to liquid water, there are fewer water molecules in the solid phase than the same volume in the liquid phase. And consequently, solid water is less dense than liquid water.

It is highly unusual for a compound's solid phase to have a lower density than the liquid phase. Why does this matter? Well, imagine if the opposite was the case, that the solid state of water had a higher density than the liquid state—if ice did not float on liquid water. Every winter, when temperatures drop below freezing, lakes and rivers would freeze from the bottom up, and life that happened to live in these bodies of water would struggle to survive. A liquid medium provides mobility not possible in a solid medium. Imagine a single-cell organism floating around in liquid water. What happens if this water is now solid? Let's assume that the water in the organism itself does not freeze (organisms that survive in freezing conditions have adaptations that prevent the aqueous cellular fluid from freezing). The organism relies on moving around to get its food, which is not possible in frozen water.

While the lower density of solid water compared to liquid allows aquatic life to survive through winters, this is also why pipes burst when the water that flows through them freezes. As water freezes, it expands, and pipes burst. In February 2021, this was the experience of millions of people in Texas when an unprecedented cold weather condition resulted in freezing temperatures and snow.[10] Freezing temperatures had been assumed to be highly unlikely in this region. As a result, water piping infrastructure did not have adequate protective measures to prevent the water flowing through these pipes from freezing.

WATER—THE SOLVENT OF LIFE

Water has another distinctive property: it is a universal solvent. A solvent is a medium in which compounds dissolve. When you add sugar to a cup of water, the sugar dissolves in water. The sugar is the solute, and water is the solvent. What happens at the molecular level when sugar dissolves in water? Sugar molecules are polar, and hence sugar molecules interact with water molecules in ways similar to water molecules interacting with each other. When opposing polar regions of sugar molecules and water molecules orient toward each other, the molecules experience attractive interactions; so, dissolving sugar results from polar water molecules interacting with polar sugar molecules. The higher the molecule's polarity, the more soluble it is in water.

The polarity of the solvent influences the kinds of compounds that dissolve in it. Because water is a highly polar molecule, it dissolves a wide range of molecules and is considered a "universal solvent." This ability to dissolve a wide range of molecules is key to why liquid water is an essential ingredient of life. Once molecules are dissolved in liquid, being in a fluid permits mobility to interact and react. Going back to early Earth before life, once conditions prevailed to allow for liquid water, the stage was set for the next phase of Earth—life. Carbon-based molecules dissolved in the liquid medium of water interacted and reacted in ways that set the stage for the chemical steps, ultimately leading to life on Earth.

The polarity of water also influences the chemistry of life. For example, the three-dimensional shapes of proteins are influenced by the aqueous environment of cells. The three-dimensional shapes of proteins support highly specialized chemical processes that have made life on Earth so complex and diverse. It is hard to imagine life demonstrating this complexity and diversity without the polar interactions made possible by a polar solvent medium like water. While ammonia is polar, it is not as polar as water and cannot dissolve as wide a range of compounds as water. Further, ammonia is a liquid at temperatures between −33°C and −78°C. At such low temperatures, the mobility of molecules is much slower than at the temperatures at which water is liquid. As a result, the speed of any chemical reaction at the temperatures that ammonia is a liquid will be significantly slower than the speed of the same chemical reaction in liquid water. The significance of the much slower rates of chemical reactions is that even if liquid ammonia could support life, it is doubtful that life in ammonia would display the complexity and diversity that life on Earth displays. Going back to Titan and its methane rain showers, methane is nonpolar. Life on Earth relies on complex interactions made possible by the polar medium of water. Similar complex biochemical interactions would not be possible in a nonpolar medium like methane. And, the frigid temperatures at which methane is a liquid mean much, much slower chemical reactions.

While water's universal solvent property makes it essential for life, solubility in water is not always helpful. For example, cell membranes—that define the boundary of a cell—are made up of phospholipid molecules. If phospholipid molecules readily dissolved in water, that would be the end of the cell. Here lies the beauty of the interplay of molecular-level interactions. Phospholipid molecules have both a polar end and a nonpolar chain (see figure 5). In an aqueous environment, the phospholipid molecules organize—self-assemble—into closed bilayer structures, as shown in figure 5. In this structure, the polar ends interact with the polar water molecules, both interior and exterior of the cell. The membrane—the cell wall—is established through interactions between the nonpolar chains lining up—like spaghetti in a box—away from the polar water and not dissolving in the water, providing structural integrity and defining the

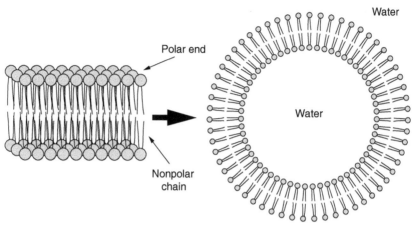

Figure 5. Self-assembly of phospholipid molecules to form a bilayer that defines the cell membrane.

Illustration by Ryann Abunuwara

boundary of the cell. The result is a well-defined cell membrane with an aqueous interior to support the chemistry of life. Simultaneously, since the cell membrane's exterior is polar, the cell can remain suspended in an aqueous environment to survive.

The two hydrogen atoms and one oxygen atom that make up the water molecule H_2O define water's properties that allow it to be a liquid on Earth's surface. Water is an excellent solvent, capable of dissolving a range of compounds that support life and provides a polar, aqueous environment to allow for complex, specific interactions between biomolecules that sustain life. There is no other naturally occurring liquid that has all these properties. As a result, water is the liquid of life. However, as explored in the next chapter, these properties that make water essential for life also make water easily contaminated and hence a potential threat to life. Herein lies the paradox of water.

3 Water

A POTENTIAL THREAT TO LIFE

According to the United Nations Children's Fund (UNICEF), women and girls worldwide collectively spend about 200 million hours (equivalent to 22,800 years) every day collecting water for their daily needs.[1] The impacts of this daily chore on women and girls include the loss of educational and economic opportunities, time spent with family, stress, and physical harm. But the tragic reality is that this water collected through such hardship may, in fact, not be safe for consumption. Moreover, even if safe at the source of collection, this water will likely become contaminated between collection and consumption.[2] Globally, over 1.23 million people died in 2019 because of unsafe water.[3]

The paradox of water is that the properties that make it essential for life also allow it to get easily contaminated and potentially threaten life. As later chapters explore, social, educational, and economic development is possible only when people have access to safe drinking water. But the chemistry of water makes this access challenging. This chapter looks at why water is so easily contaminated. By recognizing this paradox of water, we can begin to appreciate the complexities inherent in defining policies, regulations, and management practices to ensure the delivery of safe drinking water. Further, recognizing this chemistry can help inform solutions across scales—from large, centralized, infrastructure-intensive drinking water systems to community-based treatments—toward addressing the global challenge around access to safe drinking water.

Bodies of fresh water, such as lakes, rivers, and groundwater, are used as sources for drinking water. These bodies of water interact with the air and soil. As a result, natural or human activities that impact air and soil quality also affect water quality. Torrential rainfall leads to floods, which drive soil into water sources. Volcanic eruptions, forest fires, power plants, industries, and automobiles release chemicals like sulfur dioxide and nitrogen oxides into the air. These chemicals wash out in the rain and may deposit into a lake or river. Runoff from farms, including pesticides, fertilizers, and livestock waste, add chemicals and microorganisms into water sources. Similarly, rain and snowmelt runoff from streets can transfer chemicals and microorganisms into water sources. Wastewater treatment plants and industries release effluents into local bodies of water. Oil and gas extraction add to the list of contaminating sources. The point is that there is no dearth of contamination sources.

The preceding chapter looked at the distinctive chemical properties of water resulting from the molecule being made of two hydrogen and one oxygen atom described as H_2O. This chapter will explain why the same properties of H_2O that make it essential for life also make it susceptible to contamination. Specifically, given this book's focus on safe drinking water, we will look at two primary classes of contaminants: microorganisms and chemicals.[4]

LIFE THRIVES IN WATER: MICROORGANISM CONTAMINATION

As the term suggests, microorganisms, including bacteria and protozoa, are microscopic and invisible to our eyes. There are a trillion species of microorganisms that inhabit the soil, air, and water.[5] The vast majority of these organisms do not pose a threat to human health, and we rely on them (for example, bacteria in humans assist in the digestion of food and production of vitamins). As we will see in later chapters of this book, microorganisms play a crucial role in breaking down chemicals that may

Table 3 Examples of pathogens by type and disease transmitted through water

	Pathogen	Disease
Bacteria		
	Escherichia coli	Gastroenteritis
	Legionella spp.	Respiratory illness
	Salmonella Typhi	Typhoid fever
	Shigella	Dysentery
	Vibrio cholerae	Cholera
Viruses		
	Adenoviridae	Gastroenteritis, respiratory illness, eye infections
	Astroviridae	Gastroenteritis
	Hepeviridae	Infectious hepatitis
Protozoa		
	Cryptosporidium	Gastroenteritis
	Giardia	Gastroenteritis

NOTE: Viruses are not living and hence, strictly speaking, not microorganisms. But viruses can be harmful and are classified as a pathogen.

otherwise end up in our water supplies. However, a minority of microorganisms, referred to as pathogens, can be harmful to humans and may even lead to death.

Since life thrives in water, an aqueous environment will host living organisms of all sizes—from mammals, birds, fish, and plants down to microorganisms, including pathogens. Table 3 lists examples of some commonly known pathogenic microorganisms, or pathogens. The primary source of these pathogens is fecal matter from humans and animals, often due to poor sanitation practices. Runoff from livestock farms and untreated sewage are other sources of pathogens. The most common manifestation of diseases caused by these pathogens is diarrhea. Globally, about 800,000 people die from diarrheal diseases due to unsafe water and sanitation every year.[6]

Cholera, transmitted by the bacterium *Vibrio cholera*, is a classic example demonstrating the ease of the spread of waterborne diseases. Chapter 4 explores how understanding the spread of this disease through water

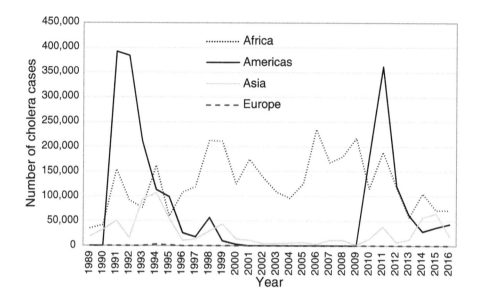

Figure 6. Data from the World Health Organization (WHO) showing global trends in reported cholera cases from 1989 to 2016.

SOURCE: World Health Organization Global Health Observatory, "Number of Reported Cases of Cholera," https://www.who.int/data/gho/data/indicators/indicator-details/GHO/number-of-reported-cases-of-cholera.

resulted in investments in drinking water treatments. Here, we will look at more recent cholera-related incidents to demonstrate why vigilance is necessary to protect water from pathogens.

Figure 6 shows the trends in the number of reported cholera cases globally.[7] *Vibrio cholera* typically enters drinking water sources due to poor sanitation practices and untreated sewage. The spike of cases in 1991 resulted from a cholera outbreak in Peru. The suspected cause was untreated sewage water released into rivers and oceans.[8] The peak in 2010–2011 depicts cholera cases in Haiti in the aftermath of the earthquake in January 2010. Cholera had been absent from this region for about 100 years.[9] In October 2010, Haiti reported an increasing number of cases of cholera. The country was already reeling from damaged infrastructure because of the earthquake. The source of the cholera was negligent sanitation practices and the release of human waste by United Nations peace-keeping troops into the Artibonite River used by Haitians for their drinking water supply.[10,11] The international troops happened to be from a region where cholera was endemic, and as a result, their waste contained *V. cholera*. Over 800,000 people were infected by cholera, with about 10,000 fatalities. Once the bacteria contaminated the river, the disease spread quickly in a region already struggling with weakened infrastructure.

While the incidence of cholera in Haiti is a tragic example of mismanagement in a stressed situation, it does not mean that waterborne diseases that impact humans are always a result of such stresses. Pathogens in drinking water sources can come from agricultural and urban runoff and human, livestock, and wildlife waste.[12] In 1993, Milwaukee, Minnesota, faced one of the largest pathogenic contaminations in drinking water in the United States. The protozoan *Cryptosporidium parvum* contaminated drinking water. This incident caused 400,000 people to fall ill and about 69 deaths. While initially, runoff from a cattle farm was the suspected source of contamination, later studies suggest that the cause may have been untreated sewage released into the local drinking water source. This water then entered the drinking water treatment, but the methods used did not address *Cryptosporidium*.[13] Researchers concluded that the total cost of this contamination was about $96.2 million, which included $31.7 million attributed to medical expenses and $64.6 million to loss of productivity.[14] In 2010, 27,000 residents (about 45%) in the town of Östersund, Sweden, were made ill by contamination of drinking water by *Cryptosporidium*.[15] In 2013, Baker City, Oregon, was impacted by the presence of *Cryptosporidium* in its drinking water supply. Cattle feces was the source of the protozoan, as cattle were allowed to graze along the edges of the source of the town's drinking water supply.[16]

The data in figure 7 reveal trends in drinking water contamination events in the United States from 1971 to 2014.[17] While there appears to be some decline in contamination events over these years, it is also evident that contamination events persist, with contamination from microorganisms at the highest frequency. These data illustrate the need for constant oversight and monitoring to keep municipal water systems free from pathogens.

Pathogenic contamination can be severe and even fatal. People consuming water contaminated by pathogens often feel the impact immediately as gastrointestinal or respiratory infections. Public health officials

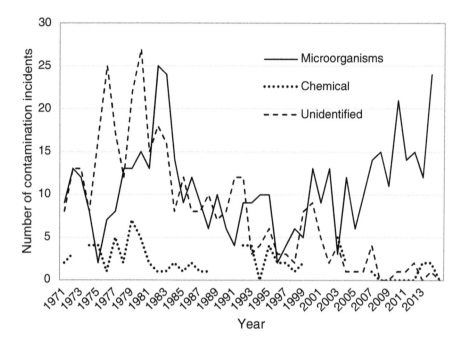

Figure 7. Data from the Centers for Disease Control and Prevention (CDC) showing trends in reported drinking water contaminations in municipal systems in the United States from 1971 to 2014. (The x-axis denotes odd years only and as a result 2014 does not appear on the axis.)

SOURCE: Centers for Disease Control and Prevention, "Etiology of 928 Drinking Water-Associated Outbreaks, by Year—United States, 1971–2014," https://www.cdc.gov/healthywater/surveillance/pdf/DW_Historical_Figure-H.pdf.

are informed when hospitals begin to note infection cases that seem higher than usual. Once the source of contamination is identified, drinking water facilities respond by utilizing treatments to address the pathogen. Local officials inform the public to boil drinking water.

For example, in February 2021, under the grip of unprecedented cold temperatures, the electrical energy system in Texas failed and resulted in power cuts to millions across the state.[18] Drinking water utilities were unable to treat water adequately. As a result, residents in many parts of Texas received "boil alerts"—boil water to kill pathogens that may be present in the water before drinking.[19]

The need for an urgent, effective response highlights the role of expertise and funding to know when and how to respond. Low-socioeconomic communities often struggle to address pathogenic contamination and resort to purchasing bottled water, which adds costs for financially strapped families.[20] One such town is Jackson, Mississippi, a predominantly Black community, where boil alerts are not uncommon.[21,22] The town has experienced declining tax revenues, which, as discussed in later chapters, impacts the ability of local municipalities to manage drinking water systems.[23] The winter storm that swept through southern states in the United States in early 2021 also impacted this community. In Jackson, delivery of tap water to residents was disrupted for weeks after the storm had passed.[24]

WATER IS THE UNIVERSAL SOLVENT: CHEMICAL CONTAMINANTS

While contamination from microorganisms is a significant concern for drinking water, contamination by chemicals is trickier to identify. Unless the drinking water delivered to homes changes in appearance, taste, or color, it is difficult to discern the presence of dissolved contaminants. Rarely are there immediate health reactions from water that may be contaminated by harmful chemicals. It may take years for health impacts to be recognized and are often realized too late. For example, residents of Hoosick Falls, New York, consumed drinking water contaminated by perfluorinated alkyl substances (PFAS) for many months before being notified of the presence of these compounds in their drinking water.[25,26] Residents in Flint, Michigan, and Newark, New Jersey, were exposed to unsafe levels of lead in their drinking water for extended periods.[27,28,29] Chapter 6 discusses these contamination events in more detail.

Being the universal solvent, water does not "discriminate" which compounds it dissolves. However, not all compounds dissolve to the same extent. For example, table salt (sodium chloride) dissolves readily in water, as does sugar (sucrose). On the other hand, calcium carbonate, which makes up shells and skeletons of coral, is said to be "sparingly" soluble in water, clearly necessary for marine organisms that rely on a shell or skeleton. At the macroscopic level, it certainly appears that a shell made of calcium carbonate stays the same when in contact with water. In contrast, we can visibly see the table salt or sugar crystals disappearing into the water. So, what is happening on the molecular scale that dictates the solubility of a compound in water?

In chapter 2, we looked at why sugar dissolves in water. Sugar molecules, being polar, interact with the polar water molecules, allowing the sugar to dissolve. What makes water the universal solvent is that its high polarity allows it to dissolve a wide range of compounds, both polar and nonpolar.

To illustrate the importance of water's ability to dissolve nonpolar compounds, let's look at the example of the oxygen molecule, O_2, and how it interacts with water. Oxygen is essential for organisms that rely on an oxygen-based, or aerobic, metabolism. Aerobic organisms include bacteria, aquatic species like fish, and mammals, including humans.

Oxygen is a nonpolar molecule but dissolves to some extent in water. Water, being a highly polar molecule, induces a temporary polarity in nonpolar molecules. This induced polarity is similar to when a magnet is rubbed against a stainless steel sewing needle. The needle becomes magnetic as a result of induced magnetism. In the case of the oxygen in water, the polar water molecules induce a temporary polarity in the oxygen molecules. As a result of this temporary polarization, there are weak attractive interactions between the oxygen and water molecules, resulting in some oxygen dissolving in water. The solubility of oxygen in water is low—*but it is not zero*. The polarization of the oxygen molecule is weak and not permanent. If the oxygen escapes from the water, the oxygen will revert to being nonpolar. However, while in an environment surrounded by water molecules, enough oxygen molecules dissolve to support the aerobic respiration of aquatic organisms like fish.

The level of dissolved oxygen is a measure of the health of lakes, rivers, ponds, and oceans in sustaining aerobic aquatic organisms. Bodies of water with low oxygen levels may not support aquatic life, resulting in "dead zones." A common cause of these dead zones is nutrient runoff, like nitrate and phosphate fertilizers, into rivers, lakes, and oceans. Excessive nutrients in water trigger algal blooms, which, when they decay, deplete oxygen levels in the water and asphyxiate other aquatic life. It is worth pausing here and recognizing that contamination of water has consequences for all life on Earth.

While water exhibits these amazing solvent properties in dissolving polar and nonpolar compounds important for life, its ability to do so is a double-edged sword. As a universal solvent, water may also dissolve compounds toxic for life. Take, for example, dioxin, 2,3,7,8-tetrachlorodibenzo-p-dioxin (TCDD). This compound has very low solubility in water. At room temperature, the maximum amount of TCDD that dissolves in one liter of water is 0.2 micrograms. To put this in context, the average mass of a grain of sugar is 0.0006 grams, or 600 micrograms. So, a solubility of 0.2 micrograms (mcg) per liter (L) for dioxin translates to a fraction of a grain of sugar invisible to the naked eye dissolved in a liter of water. Compared to TCDD, sugar is millions of times more soluble in water. If TCDD has such a low solubility in water, why is this chemical discussed here?

Burning carbon-based fuels like wood, coal, and oil and incinerating municipal and industrial waste can release TCDD. The health effects of TCDD, even at low concentrations, include an increased risk of cancers and reproductive difficulties. TCDD has been detected in drinking water sources. Even though it is not very soluble in water, its solubility is not zero. TCDD levels exceeding 0.00003 mcg/L in water are considered unsafe for humans. TCDD can dissolve to a concentration that exceeds its safe level by 6,700 times. As we will see in chapter 5, TCDD is just one on a long list of chemicals detected in drinking water sources that are potential health risks for humans. The list is long because water is the universal solvent and dissolves compounds it comes into contact with. The example of TCDD shows that even if the compound does not have a high solubility in water, it dissolves enough to be a health risk.

Another solvent property of water is its ability to dissolve ionic compounds. One example of an ionic compound is table salt, sodium chloride (NaCl), which is made up of a positively charged sodium ion (Na^+) and a negatively charged chloride ion (Cl^-). In NaCl, the sodium and chloride ions are oppositely charged and experience attractive interactions. Yet, water dissolves sodium chloride since H_2O can interact with both the sodium ion and chloride ion. Once again, the high polarity of water allows it to dissolve ionic compounds like sodium chloride.

Life takes advantage of the fact that compounds like sodium chloride dissolve in water. Cellular processes rely on differences in concentrations of ions inside and outside the cell, referred to as a "concentration gradient" across the cell membrane. Such cellular processes include nerve responses, muscle contraction, and hormone secretion. Water's ability to dissolve ionic compounds and sustain ions in an aqueous environment supports life's complex biochemistry. Regulating this cellular concentration gradient of ions is so critical that drinking large quantities of distilled water may be dangerous. Distilled water is essentially pure water with no dissolved ions, that is, no mineral content. Consuming large amounts of distilled water may affect this concentration gradient, leading to potentially harmful cell death. Hence, despite 70 percent of Earth being covered with water, the fact that this water is salty prevents us from consuming large quantities of ocean water, as doing so will affect this concentration gradient. For oceans to be a source of drinking water for humans, this water must be treated through a process called desalination. As the term suggests, this treatment lowers the level of dissolved salts, which are the dissolved ions. For similar reasons, dehydration can be dangerous if this concentration gradient of ions is affected.

If you take a sip of distilled water, it will likely taste flat, as our saliva contains dissolved ions. Water that tastes "right" has similar ionic content as our saliva. Tap water, even after treatment, contains some dissolved ions and hence has some mineral content. Some bottled water companies that use filtered tap water as the source add some ions postfiltering to make the water more appealing. Water filtered through home-based filters that attest to removing all dissolved minerals may not necessarily have a taste you enjoy or be optimal for your cellular functioning if consumed in large quantities.

Since water dissolves ions, it once again does so indiscriminately. As a result, water dissolves lead, nitrate, and arsenic ions, all of which have harmful health impacts. The use of lead in pipes and plumbing resulted from building practices in the 1900s, and many cities still rely on these old pipes. These plumbing fixtures are often the source of lead ions in drinking water. Nitrate ions in drinking water sources result from agricultural runoff due to nitrate-based fertilizers. Chapter 6 discusses the history of lead and nitrate contamination in drinking water. Arsenic ions occur naturally in some soils and have industrial sources. All these ions readily dissolve in water and are harmful to human health.

Due to the fundamental nature of water, there will always be microorganisms and chemicals present in water. Water exposed to air, soil, and natural and human activity cannot be pure because of the chemistry of H_2O. Not all microorganisms and chemical compounds are harmful, and levels present in drinking water sources may not be a threat to humans. However, they may be detrimental to other organisms. The point is that pure water is not available on Earth. Life does not require pure water, but life does need *safe* water.

In many communities, local drinking water sources are inadequately protected, resulting in chemical and biological contaminations. As discussed in later chapters, an increasing number of chemicals used in industry, agriculture, pharmaceuticals, and personal care products are detected in natural bodies of water, some of which are sources for drinking water. We have insufficient assessments of the impacts of exposure on human and ecosystem health for some of these chemicals. As a result, there is an urgent need for stringent regulations, oversight, and investments to address existing and new threats to ensure the delivery of safe drinking water. Moreover, once water is contaminated, rendering it safe is a complicated and expensive process, as it is not trivial to extract dissolved chemicals and deactivate pathogens that may be present.

The fundamental chemistry of water makes the delivery of safe water challenging, requires expertise and investments, and demands constant vigilance. These investments are recovered from the social and economic gains possible when communities have access to safe drinking water. We must honor and respect water for the life-giving paradoxical medium it is.

4 Why Drinking Water Quality Matters

Geologists refer to the past 11,700 years as the Holocene epoch, which signifies the end of the last ice age on Earth. Over this epoch, Earth has experienced relatively warmer average global surface temperatures and climate stability for an extended period. Scientists believe that this relative climate stability is a reason for humans' transition from a nomadic, hunter-gatherer lifestyle to a settled lifestyle and the establishment of modern civilizations. A shift away from a nomadic lifestyle required sustained access to water sources and water storage and management strategies.

Archaeological remains from 10,000 years ago in San Marcos Necoxtla, Mexico, and from 8300 BCE in Cyprus reveal that water-storing strategies evolved thousands of years ago.[1] Structures dating from before 7500 BCE in modern-day south Jordan have been identified as cisterns to capture water.[2] These archaeological remains are evidence that our ancestors recognized that sustained access to water was essential for their survival. The great civilizations of the past were on the banks of rivers—the Tigris and Euphrates, Nile, Indus, and Hueng He. Being situated on riverbanks allowed continued access to water for daily needs and establishing farming and maintaining livestock. Agriculture and livestock translated to control over food, which over time supported population growth.

While global temperatures were relatively stable, local variations in the climate impacted communities. An ancient site in the Beidha area (near Petra, Jordan) reveals a decline in human presence around 13,000 years ago when the region experienced drought.[3] Around 9500 BCE, this region was wetter, and humans repopulated the site. Then, around 6500 BCE, the area was again abandoned, and evidence suggests this region experienced another drought. Archaeologists believe that the climate in the Indus Valley changed about 3900–3000 years ago, leading to extended droughts, which may have triggered the collapse of this civilization.[4] People in the Indus Valley moved further east and settled along the Ganges, where precipitation patterns were more stable. The collapse of the Mayan civilization may have also been influenced by changes in the local climate, which caused many severe droughts, each lasting between 3 to 18 years.[5] These examples highlight the crucial role sustained access to water plays in supporting the survival of a civilization.

Recognizing the need for water, communities established practices and enacted laws that respected and honored the right to access to water. For example, according to the Talmud, "rivers and streams forming springs, these belong to every man."[6,7,8] The Koran recognized that "anyone who gives water to a living creature will be rewarded. To the man who refuses his surplus water, Allah will say: 'Today I refuse thee my favor, just as thou refused the surplus of something that thou hadst not made thyself.'"[9] Many indigenous communities continue to uphold this right to water and that water is a shared asset.[10] The Bedouin and Berbers believe that "water to quench thirst is an unalienable right and may not be refused from any water source."[11]

While access to water is essential, the quality of this water is also crucial. When did humans recognize the importance of water quality? Archaeological evidence from 4000 BCE reveals that humans were concerned about the aesthetic quality of water as measured by appearance, taste, and odor.[12] Ancient Sanskrit and Greek writings describe methods to improve water's aesthetic quality by straining, filtration with sand and charcoal, and exposure to sunlight. Records from Egypt from 1500 BCE indicate the use of alum to lower turbidity, a measure of the level of suspended particles in water. Humans used visual identifications like flowing water or clear appearance to decide if a water source was suitable for drinking.[13,14] Unfortunately, as we will see later, these physical characteristics of appearance, taste, odor, and color are insufficient indicators of water quality and safety.

IMPACTS ON DRINKING WATER QUALITY FROM POPULATION GROWTH

As humans settled and populations expanded, the problem of human and animal waste disposal arose. Ruins from civilizations in Mesopotamia and the Indus Valley, and later during the Greek and Roman empires, reveal toilets with sewer systems that removed the waste.[15] The sewer systems dumped this waste into local bodies of water, which were also the source of water for the community. As communities grew and the amount of waste dumping in local water sources increased, the aesthetic measures used to assess water quality—odor, taste, appearance—deemed the water unfit for human consumption. One way of addressing contamination of local water sources is to bring water from more pristine areas. The ancient Romans took this approach and embarked on massive engineering projects, including constructing aqueducts to bring water from remote regions into cities. The Romans also built the Cloaca Maxima, a sewer system, to remove waste from toilets, baths, and runoff from the streets. The sewer system emptied into the local water sources, such as the Tiber River.[16,17]

Population growth over the Medieval Ages and the Industrial Revolution exacerbated two water-related challenges: growing demand for water and increasing pollution of local water sources. Cities in Europe resorted to the ancient Roman method of moving water from remote, less polluted water sources into cities. Private water carriers brought water into cities and sold the water to residents. Public funds assisted in constructing pipes that brought water to fountains within cities and where locals collected water. Bringing water to the cities may have addressed demand and provided better quality water, but this did not address the increasing amounts of pollution in cities from human and domestic animal waste and industrial activities such as tanneries and slaughterhouses—the stench of waste engulfed cities.

Figure 8. Cartoons that appeared in the 1850s in *Punch*, a British magazine, highlighted public sentiment about the quality of the water in the Thames River. Left: "A Drop of London Water," 1850. Right: "The Silent Highway Man," 1858.

SOURCE: "A Drop of London Water," *Punch*, May 11, 1850, 188. "The Silent Highway Man," *Punch*, July 10, 1858, 137.

Plagues such as bubonic, cholera, and typhoid fever were frequent during these eras. For example, cholera outbreaks devastated London in 1832, 1849, and 1855, causing tens of thousands of deaths.[18,19] At this point, people believed the "miasma theory" that diseases spread through the vapors emanating from the waste and permeating the air. As a result of the acceptance of the miasma theory, cities washed away the stench by using water to clean streets. Outlets of sewer systems constructed farther downstream also helped get rid of the smells. All this cleanup effort just added to the contamination of local bodies of water and impacted residents downstream of the sewer outlets. In a series of cartoons, *Punch*, a weekly British magazine of this period, captured the public sentiment in London about the water quality in the Thames River (figure 8).

In London, the stench from the growing pollution of the Thames River peaked in 1858—dubbed the "year of the Great Stink." It was a warm summer, and decades of dumping resulted in a stench that engulfed the city. The unbearable smell finally forced the city politicians to address the source—the rotting waste in the Thames—and raise the necessary funds to address this problem. Joseph Bazalgette, hired as the city engineer, established the infrastructure for an underground sewer system that released the city's wastewater farther downstream.[20,21] The sewer system addressed the stench, and over time the quality of the water in the Thames improved. Cholera rates declined. Addressing the stench appeared to successfully halt the spread of infectious diseases, a seeming victory for the miasma theory. However, an outbreak of cholera in 1866 in the East End region of London raised questions about the validity of the miasma theory, allowing the "germ theory" to be accepted as the model for transmission of infectious diseases.

The ancient Greeks were the first to raise the idea that diseases were spread by a "contagion" or "seeds." However, the nature of this contagion was not understood. As a result, the miasma theory prevailed through the Middle Ages and the Industrial Revolution. The miasma theory made sense to people as higher disease rates were in cities where unpleasant odors persisted compared to the countryside. In the 1670s, Anton van Leeuwenhoek was one of the first people to observe microscopic life in water and soil samples. At this point, the term "animalcules" was coined to refer to these organisms. Scientists proposed that some of these animalcules might be responsible for spreading diseases. But, the miasma theory prevailed as the accepted model of disease transmission. Odors trigger our sense of smell, whereas microorganisms are invisible to the naked eye. This invisibility of microorganisms made it difficult for people to accept them as the causative agents of diseases.

In 1854, John Snow's meticulous work identified a cholera epidemic's source as a specific water pump located on Broad Street in London. He demonstrated that the disease spread through the water and not the air. While Snow presented his results in 1854, predating the Great Stink by four years, his theory of disease spread through water was not accepted. And it did appear initially that Bazalgette's sewage system had quenched the spread of cholera. However, the 1866 cholera epidemic finally raised questions that the miasma theory could not answer. A leak from the sewer system into the water distribution pipes in this region contaminated the water. Residents consuming this contaminated water contracted cholera. It was clear that the spread in this region of London was through the water and not the air.

The 1866 East End cholera epidemic overlapped with work by scientists such as Louis Pasteur and Robert Koch. Pasteur's and Koch's research, along with the work of other scientists, established the "germ theory," which states that microorganisms can spread diseases through contaminated water or food. In 1883, Robert Koch isolated the bacterium *Vibrio cholera*. He demonstrated that cholera is a waterborne disease spread by consuming water or food contaminated with *V. cholera* and that a source of this bacteria is human waste. The work of scientists, including John Snow, Robert Koch, Louis Pasteur, Karl Joseph Eberth, and Georg Theodor August Gaffky, in identifying biological vectors of diseases such as cholera and typhoid fever enabled the acceptance of the germ theory. Scientists and health practitioners now understood that the spread of waterborne diseases was due to microorganisms and not odors in the air. Studies also demonstrated that these microorganisms spread due to poor sanitation practices and the dumping of human and animal waste into local bodies of water. Those who consumed this water were infected, and their waste was added to the pathogenic load in water sources.

PUBLIC HEALTH BENEFITS OF DRINKING WATER TREATMENT

During the 1800s, just as cities in Europe were plagued by unsanitary conditions and dumping waste into local bodies of water, burgeoning cities in the United States were dealing with similar challenges. New York, Chicago, and Philadelphia had epidemics of cholera and typhoid. With the acceptance of the germ theory, scientists and engineers in Europe and the United States researched ways to remove pathogens from drinking water. Sand filtration, a method used since ancient periods, demonstrably lowered levels of pathogens. Researchers also investigated disinfection methods for killing pathogens. The effectiveness of sunlight, boiling, ozone, copper, silver, and chlorination was tested. Ultimately researchers determined that chlorination was an optimum choice for killing pathogens and the cheapest for large-scale treatment. The late 19th and early 20th centuries saw the establishment of drinking water treatment facilities where sand filtration, along with chlorine disinfection, became the standard approach. The treated water was delivered through pipes to consumers.

Before the wide-scale adoption of drinking water and sewage treatment, about 25 percent of reported deaths were due to waterborne infectious diseases.[22] The mortality rates for people living in urban cities were higher than for people living in rural communities. Poor air and water quality and inadequate sanitation in cities contributed to high mortality rates. Cities that instituted drinking water treatments saw a decline in mortality rates due to waterborne diseases such as typhoid fever.[23,24,25,26] Life expectancy increased at a rate higher than what could be attributed only to the decrease in deaths due to typhoid fever. The recognition of the impact of drinking water quality on increasing life expectancy is captured in a quote in 1909 by Alan Hazen, who researched water filtration systems: "Where one death from typhoid fever has been avoided by the use of better water, a certain number of deaths, probably two or three, from other causes have been avoided."[27] Along with antibiotics and vaccines, the establishment of drinking water treatment is heralded as a significant public health success of the 20th century. The public, which remembered the effect of water that made them ill, or worse, killed family members, cherished this access, as captured in a 1919 cartoon in *The American City* (figure 9).

Ironically, while cities invested in drinking water infrastructure in the early 20th century, treating wastewater before releasing it into rivers and other bodies of water was deemed an unnecessary expense.[28,29,30] The visible contamination of rivers and local bodies of water triggered some politicians to raise awareness of this water pollution. Yet, it was a hard sell to get cities to pay for wastewater treatment, particularly with other, more pressing, social, political, and economic challenges of the early 20th century. In the United States, it was not until the late 1940s when public outcry over the stench from the polluted bodies of water triggered the 1948 Water Pollution Control Act.[31] The final step toward regulations for drinking water and wastewater was the establishment of the Environmental Protection Agency (EPA) in 1970 and the passing of the 1972 Clean Water Act and the 1974 Safe Drinking Water Act (chapter 5 discusses these acts).

As a result of research identifying the causative agents of the spread of waterborne diseases, the establishment of drinking water treatments, government investments in the necessary infrastructure, and environmental

Figure 9. A cartoon that appeared in a 1919 issue of *The American City* championing chlorine disinfection for treating drinking water.

SOURCE: Cartoon by Eugene Zimmerman, *The American City* 21 (September 1919): 247.

regulations, today the majority of residents in most Global North nations have access to safe drinking water out of their taps. As mentioned in chapter 1, the World Health Organization (WHO) defines safe drinking water as water that "does not represent any significant risk to health over a lifetime of consumption, including different sensitivities that may occur between life stages." Chapter 5 will look more closely at the processes involved in treating water from source to taps. However, it is worth pausing and recognizing what it means when a nation can provide seemingly unlimited and safe water to homes, schools, and businesses. Access to safe drinking water means fewer sick days for children and adults. Children have healthier lives and more days at school, and adults can increase their economic productivity. Establishing the infrastructure for treating drinking water and wastewater was crucial in allowing people who benefited from these public health initiatives to "grow old."[32]

GLOBAL ACCESS TO SAFE DRINKING WATER

Across the globe, it is estimated that about two billion people lack access to safe drinking water.[33] Young girls in many rural communities often walk many miles a day to collect and carry water back to their homes. These trips can be dangerous. Further, the time spent collecting water means that these girls do not have the time to get an education. According to UNICEF, the amount of time spent daily by women and girls to gather water is 200 million hours (equivalent to 8.3 million days, or 22,800 years).[34] Every year, 443 million school days are lost to water-related illnesses.[35] A child under five dies every two minutes, and a newborn dies every minute due to a lack of safe water and poor sanitation.[36] These are sobering statistics.

Why do so many people lack access to safe drinking water when the knowledge and technology for ensuring this has existed for over a century? The late 19th and 20th centuries was the era of large infrastructure projects supporting the delivery of safe water and sewage treatment in Europe and the United States. But what about parts of the world that are currently challenged by these very same issues? The answer, in large part, is colonialism. A worthwhile expenditure for citizens at home was not deemed equivalently worthy for people of the "colonies."[37] After gaining independence, many of these former colonies were left to figure out how to govern virtually overnight and often without financial means. Many of us do not appreciate that the infrastructure for drinking water and wastewater treatment is expensive and complicated and requires regulatory frameworks, constant oversight, and funding. For many countries, these funds were just not available and that continues to be the case. In some places, political instability makes it challenging to establish the infrastructure and meet the intensive regulatory demands of safe drinking water and wastewater treatment. Even in nations that have established environmental regulations, including ones in the Global North, political parties influence how these regulations are enforced.

SUSTAINABLE DEVELOPMENT AND ACCESS TO SAFE DRINKING WATER

Recognition of the challenges faced by nations, many of which struggled after colonialism, led to the United Nations' Millennium Development Goals (MDGs).[38] Adopted in 2000, the UN Millennium Declaration "affirmed Member States' faith in the United Nations and its Charter as indispensable for a more peaceful, prosperous, and just world. The collective responsibility of the governments of the world to uphold human dignity, equality and equity is recognized, as is the duty of world leaders to all people, and especially children and the most vulnerable."[39] To realize this global responsibility, the Declaration defined eight MDGs to be achieved by 2015. Goal number seven of the MDGs focused on environmental sustainability, which included the target "Halve, by 2015, the population without sustainable access to safe drinking water and sanitation." As a result of global efforts, by 2015, the target for safe drinking water was exceeded, however, not for sanitation.[40]

In 2015, to continue the progress achieved through the MDGs, the UN's member states adopted the 2030 Sustainable Development Goals (SDGs). An essential declaration of the SDGs is a global resolve "to end poverty and hunger everywhere; to combat inequalities within and among countries; to build peaceful, just and inclusive societies; to protect human rights and

Table 4 The Joint Monitoring Programme (JMP) "drinking water ladder" classifications, based on access and water quality

Label	Definition
Safely managed	Drinking water from an improved water source which is located on-premises, available when needed, and free from fecal and priority chemical contamination
Basic	Drinking water from an improved source, provided collection time is not more than 30 minutes for a round trip, including queuing
Limited	Drinking water from an improved source for which collection time exceeds 30 minutes for a round trip, including queuing
Unimproved	Drinking water from an unprotected dug well or unprotected spring
Surface water	Drinking water directly from a river, dam, lake, pond, stream, canal, or irrigation canal

SOURCE: WHO/UNICEF Joint Monitoring Programme. "Drinking Water." https://washdata.org/monitoring/drinking-water.

promote gender equality and the empowerment of women and girls, and to ensure the lasting protection of the planet and its natural resources."[41] Of the 17 SDGs, the goal of SDG 6 specifically "ensures availability and sustainable management of water and sanitation for all."[42] SDG 6 acknowledges the critical role of access to safe drinking water in the human, social, and economic development of a nation. Gains made by countries in meeting SDG 6 will have far-reaching impacts on their residents.

The WHO/UNICEF Joint Monitoring Programme for Water Supply, Sanitation and Hygiene (JMP) gathers country-level data that informs global assessments around access to safe drinking water, sanitation, and hygiene practices.[43] These data are used to monitor nations' progress in providing these services to their people and meeting the targets set by SDG 6. Table 4 describes the labels and metrics used by the JMP that classify drinking water sources based on ease of access and the quality of water.[44]

The data in figure 10 show the fraction of people in different geographic regions with access to the JMP categories of water quality.[45] As seen in figure 10, compared to regions of the Global North (North America and Europe), regions of the Global South have higher fractions of popula-

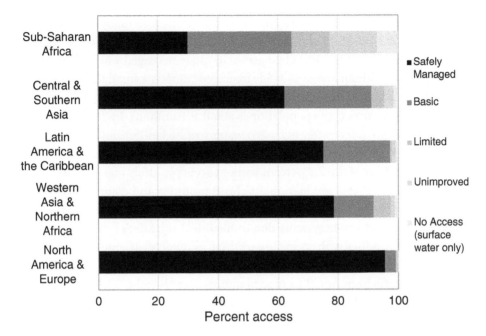

Figure 10. Graphs showing the percentage of people across geographic regions with access to the water source and quality categories defined by the Joint Monitoring Programme (JMP) drinking water ladder.

Reproduced with permission from WHO/UNICEF Joint Monitoring Programme, "Data," https://washdata.org/data.

tions that lack access to safely managed water—many nations in these regions being former colonies. It is in these regions where the majority of the two billion people who lack access to safe drinking water reside. People in countries with limited access to safely managed water are in a vicious cycle. Lack of funding and infrastructure make it challenging to provide easy access to safe water. Simultaneously, not having this access robs communities of healthier, more equitable, and economically productive lives.

But lest we assume that access to safe drinking water is a challenge only for residents of countries in the Global South, we see failures even in countries such as the United States that have instituted laws and practices for safe water. Similar to the history of colonialism that denied people in the colonies access to safe water, racism in the United States restricted access to drinking water and sanitation infrastructure to many communities of color.[46] In 2020, the JMP data for the United States indicated that 97.3 percent of its residents had access to safely managed water.[47] While 2.7 percent without access to safely managed water is a small fraction, for the United States, this translates to about 8.9 million people (as a comparison, New York City's population is about 8.4 million). In 2020, about 30 nations provided safely managed water to a higher percentage of residents than the United States, and 12 claimed 100 percent coverage.[48]

The majority of the 2.7 percent without access to safely managed water in the United States are marginalized communities and communities of color.[49,50] Many Native communities in the United States do not have access to piped water systems due to historical injustices.[51,52] People in these communities with access to cars drive as much as 40 miles to collect water in containers to bring to their homes and communities. Chemical contaminants from runoff from agriculture often taint the drinking water delivered to residents in the Central Valley of California.[53] This region is called the country's food basket as produce grown here is sent all over the United States. Yet, the industry feeding us is responsible for contaminating water sources and making the water unsafe for the communities who work on these farms. In Appalachia, toxic metals such as arsenic and mercury leach from coal mines and coal ash into drinking water sources. People who work in this hazardous industry are also exposed to toxic chemicals in their drinking water.[54]

ACCESS TO WATER IS TRANSFORMATIVE

The social and economic benefits of access to safe drinking water are significant. A 2016 study investigated the impacts on a community when provided access to water.[55] The study site was a region of Kenya that has benefited from projects giving the community access to improved water sources.[56] Before these water intervention projects, community members walked many hours a day to collect water. The following comments highlight the day-to-day challenges faced when water is unavailable locally. Such problems are faced globally by hundreds of millions of people every day.

> I used to go to Athi River for four hours and come back with 25 liters of water. Imagine 25 liters of water with a family of 10 people; you wouldn't have water to bathe, to clean the clothes, it was only for cooking and drinking.[57]
>
> I used to lose business because when I went to the river, I would have to close my shop. When they found I had closed, they would move on to other shops.[58]

Many day-to-day challenges are resolved once people have access to water locally and from improved sources. Community members discussed feeling much less stress, having time for family, addressing family members' needs, and having time for more economically productive tasks, resulting in improved quality of life. The following comments from members of this community highlight the transformative power of water:

> It [my family] has been strengthened by having the water nearby because we are able to get water and be together here. I won't be gone half a day or a whole day—so that will give us time to be together as a family when my husband comes from work, my kids from school . . . we can sit and talk about how the day was, and if there are any issues to discuss, we can talk about it. Maybe a kid goes to school, and he needs some money, we can use that opportunity to discuss it and meet those needs.[59]
>
> The animals, especially during the dry spell, used to die on the way to Athi because of their poor body condition, but now . . . they get water from the project here. Like, a goat this size—it goes for nearly 6000 [Kenyan shillings]! That is good money! Because it's healthy, you take [the animal] to the market, and you will be able to get good money. You can use for your other needs, pay school fees or buy food . . . that is from having the water project near us.[60]

While this is just one study, this community's reactions seem universal when people are under stress and experience relief once this stress is lifted. Time allocated to fetching water is now used for family, education, and socially and economically positive efforts. Imagine if every person who currently struggles to access water and walks miles a day to fetch water, which is likely unsafe, could experience what this community now enjoys and benefits from.

Once scientists established how diseases spread through water, cities responded by investing in the infrastructure to treat drinking water.

Investments made in the 20th century were recovered through improved public health gains and expanded educational, social, and economic opportunities. However, in the 21st century, many of us in the Global North take this 24/7 access to high-quality water for granted, which is dangerous given the ease with which water gets contaminated.[61] There are growing threats to the delivery of safe water, including failing infrastructure, reduced funding for drinking water treatment systems, increasing numbers of chemicals detected in drinking water sources, and climate change altering precipitation patterns and stressing freshwater sources. It is time for all of us to wake up to the fact that we cannot take access to safe drinking water for granted.

In the next chapter, using the United States as a case study, we will look at the path water takes from source to tap and the regulations that dictate the management of safe drinking water. The more each of us understands the steps from source to tap, the more informed will our demands be to our political representatives to ensure sustainable, equitable access to safe drinking water and investing in the nation's health and economic well-being.

5 Making Water Safe

Water covers about 70 percent of Earth's surface, with 97 percent in oceans and 3 percent in fresh water. Of this 3 percent fresh water, ice caps and glaciers lock 75 percent, 24.5 percent is groundwater, and 0.5 percent is in surface water sources, such as lakes, rivers, and streams, moisture in the soil, and water in living organisms (see figure 11). From a human consumption perspective, less than 1 percent of all water on Earth is available as groundwater, lakes, rivers, and streams. Furthermore, given water's chemistry, this fresh water will not be pure; there will always be microorganisms and dissolved chemicals. So, how do we assess whether this water is safe for humans?

In the late 19th and early 20th centuries, cities such as London, New York, Chicago, and Philadelphia invested in infrastructure to treat drinking water. The question then arose, how can water quality be monitored to prevent an outbreak of cholera or typhoid? To address this requires metrics and measures to assess the quality of the treated water. Over time, as cities established drinking water systems, the need for local and national level coordination became necessary to ensure consistency in standards and regulations. In other words, ideally, the quality of tap

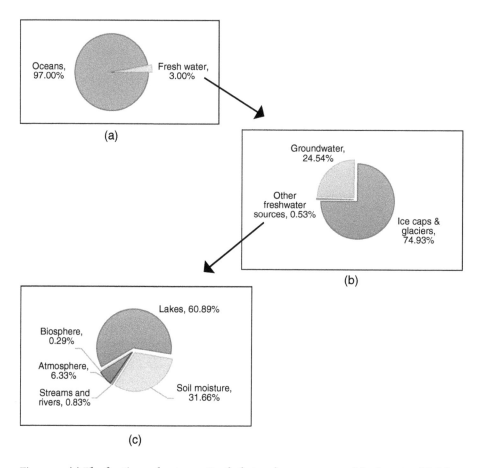

Figure 11. (a) The fractions of water on Earth that make up oceans and fresh water. (b) Of the 3% fresh water, 75% is in ice caps and glaciers, 24.5% is in groundwater, and 0.5% is in other freshwater sources. (c) The other freshwater sources include lakes, streams, and rivers.

Illustration by Ryann Abunuwara

water should not depend on your zip code (though, as we will see later, this is not the case today). Using the United States as a case study, this chapter explores how these issues were addressed. It was a long process, and, even today, it is an evolving process with successes, failures, and challenges.

REGULATING DRINKING WATER

The first regulation to safeguard human health from waterborne diseases in the United States was the 1912 abolishment of the "common cup" on interstate transportation. During this time, it was often the case for a water fountain to have a cup attached to it from which people drank water—a "common cup."[1] A statement in a 1913 annual report of the New Jersey Board of Health captures the health concerns of the use of a common cup:

> One of the representatives of this Board [New Jersey State Health Board] while traveling on a railroad train noted that a family of children [was] afflicted with whooping-cough. As the children had spasmodic attacks, after each attack had passed they would go to the water cooler and take a drink from the glass, which was used in common by all the passengers. After this had been repeated several times, the inspector took occasion to go to the cooler, and holding the glass to the light found that it was smeared with the infected mucous from the mouths of these children.[2]

In 1914, acting on a recommendation by the United States Surgeon General, the Treasury Department adopted a bacteriological standard for drinking water.[3] This standard applied to drinking water supplied by carriers (trains, buses, etc.) that crossed interstate lines. The published standard included protocols to quantify the bacterial content in water. The tests assessed if the bacteria present in the water were from the gut of warm-blooded animals, which indicates contamination from human and animal waste. Officials responsible for drinking water quality were also mindful of "unnecessary" costs associated with treating water as indicated in this quote from the standard: "Limits upon these impurities must accordingly be so placed as to allow the public an ample margin of safety, but to do this raises the question as to how far it is justifiable to tax the carriers to eliminate impurities whose deleterious effects are so doubtful."[4] Some states adopted the 1914 standards as guidelines for monitoring the quality of drinking water.

In the early 20th century, the chemical industry, established in the late 18th century, experienced a rapid expansion. Chemicals extracted from sources such as crude oil (petrochemicals) were (and still are) used as starting ingredients for many products. This era also saw the growth of the "synthetic" chemical industry to manufacture large quantities of compounds, including ones present in natural sources and many that do not have natural sources. These chemicals are used to manufacture products such as pesticides, pharmaceuticals, personal care products, paper—basically many items used daily in homes, businesses, schools, and so on.

In the 1950s and '60s, analyses of air, soil, and water samples revealed the presence of these synthetic chemicals, raising questions as to whether these chemicals impact ecosystem and human health. Rachel Carson's book *Silent Spring*, released in 1962, powerfully responded to these concerns. This book eloquently describes the damage to ecosystems as a result of our reliance on synthetic chemicals. Medical and public health researchers also cautioned that some compounds might be carcinogenic.[5,6] In 1962, in response to these concerns, the US Public Health Service revised its drinking water standards and, for the first time, listed specific chemicals and radioactive elements for screening when selecting drinking water sources.[7]

The 1960s, a time of social and political upheaval in the United States, was also when the public became aware of environmental degradation due to human activity. Air pollution was evident in the sooty smog, burning eyes, and acrid smells, which triggered respiratory illnesses such as asthma. Rivers and lakes were catching on fire. Untreated waste discharged from sewage systems, oil spills from petroleum refineries, and chemical waste released from industries contaminated bodies of water. An article in *Time* magazine in 1969 described the horrific state of rivers in the United States:

> The Potomac reaches the nation's capital as a pleasant stream and leaves it stinking from the 240 million gallons of wastes that are flushed into it daily. Among other horrors, while Omaha's meatpackers fill the Missouri River with animal grease balls as big as oranges, St. Louis takes its drinking water from the muddy lower Missouri because the Mississippi is far filthier. Scores of US rivers are severely polluted—the swift Chattahoochee, majestic Hudson, and quiet Milwaukee, plus the Buffalo, Merrimack, Monongahela, Niagara, Delaware, Rouge, Escambia and Havasupai. Among the worst of them all is the 80-mile-long Cuyahoga, which splits Cleveland as it reaches the shores of Lake Erie.[8]

While many indigenous cultures embrace "ecocentric" and "nonmaterialistic" values and ethics, these sensitivities were not (and arguably are still not) embedded into many societies and practices.[9] It was not until the 1960s that such visible evidence of environmental degradation, and books such as *Silent Spring*, triggered an ecological consciousness in the public. During this era, people saw their planet from space for the first time. Images such as *Earthrise* taken by the Apollo 8 mission in 1968 brought home the fragility of our planet. On April 22, 1970, people celebrated the first Earth Day.

In response to a public demanding action to protect the environment, the US Congress passed the National Environmental Policy Act (NEPA), signed into law in 1970 by President Richard Nixon. This law aimed to "assure that all branches of government give proper consideration to the environment prior to undertaking any major federal action that significantly affects the environment."[10] NEPA also established the Council on Environmental Quality (CEQ) as a division within the executive branch of the US government. The CEQ put forth a proposal to consolidate environmental responsibilities in different federal agencies under a new agency. By the end of 1970, Congress approved the establishment of this new agency, the Environmental Protection Agency (EPA), with the following charge:

Establish and enforce environmental protection standards.

Conduct environmental research.

Provide assistance to others combatting environmental pollution.

Develop and recommend to the President new policies for environmental protection.[11]

President Nixon's charge to the first EPA administrator, William Ruckelshaus, was to view "the environment as a whole," "treat air pollution, water pollution and solid wastes as different forms of a single problem," and introduce a "broad systems approach [that] ... would give unique direction to our war on pollution."[12]

Before 1970, while federal environmental laws such as the 1955 Air Pollution Act, 1963 Clean Air Act, and 1965 Water Quality Act were passed, enforcement was weak. As a result, compliance varied between states. The establishment of the EPA marked a shift in governance around environmental issues. The EPA is authorized to define mandatory regulations that require compliance from all states. In 1970 Congress passed the Clean Air Act, which differed from the 1963 act as now mandatory air quality standards had to be met by all states in the entire nation. In 1972 Congress passed the Federal Water Pollution Control Act, also known as the Clean Water Act (CWA). The 1972 CWA authorized the EPA to regulate water quality to "restore and maintain the chemical, physical, and biological integrity of the nation's waters."[13] The CWA established enforceable federal standards to protect surface water from becoming dumping grounds from point sources such as pipes from wastewater treatment plants and industrial effluents.[14]

In 1974, Congress passed the Safe Drinking Water Act (SDWA).[15,16] The impetus for the act was captured in a statement in a 1974 article by Peter Kyros, a congressional representative from Maine who served on the House Public Health Committee: "New pollutants have begun to degrade the quality of drinking water. The use of toxic chemical compounds for industrial, institutional, agricultural, and domestic purposes has increased tremendously. As glamorous as these new chemicals may be, they do not vanish like Cinderella at the stroke of twelve. They enter and contaminate surface and groundwater."[17]

NATIONAL PRIMARY DRINKING WATER REGULATIONS

Under the SDWA, the EPA regulates drinking water quality through the National Primary Drinking Water Regulations (NPDWR).[18] These "are legally enforceable primary standards and treatment techniques that apply to public water systems. Primary standards and treatment techniques protect public health by limiting the levels of contaminants in drinking water."[19] The primary standards limit the concentrations of 90+ regulated contaminants in drinking water.[20] Scientific, medical, and public health research inform decisions on which contaminants should be regulated and at what levels. As discussed below and in the chapters that follow, these decisions are influenced by economic and political interests. The regulations also define nonmandatory secondary standards that assess water's aesthetic aspects, such as taste, color, and odor.

Table 5 lists the different categories of drinking water contaminants, typical sources, examples of contaminants for each category, and potential

Table 5 Categories of contaminants regulated under the National Primary Drinking Water Regulations (NPDWR) and their sources, examples, and health effects

Contaminant category	Definition	Sources	Examples	Health effects
Microorganisms	Pathogens including bacteria, protozoa, and viruses	Human and animal waste	Fecal coliform and *E. coli*, *Giardia lamblia*, *Legionella*	Gastrointestinal illness (e.g., diarrhea, vomiting, cramps)
Organic chemicals	Chemicals that contain carbon atoms	Industry, agriculture, wastewater, natural	Atrazine, dioxin, benzene, styrene	Increased risk of cancer, reproductive difficulties, liver and kidney damage, neurological, delays in mental development
Inorganic chemicals	Chemicals that do not contain carbon atoms	Industry, agriculture, wastewater, natural	Lead, nitrate, cadmium, mercury, chromium	Increased risk of cancer, reproductive difficulties, liver and kidney damage, neurological, delays in mental development
Radionuclides	Radioactive elements	Natural deposits and runoff from mines	Uranium, radium	Increased risk of cancer, kidney toxicity
Disinfectants	Compounds used to disinfect drinking water from microorganisms	Process of disinfection	Chlorine, chloramines, and chlorine dioxide	Eye/nose irritation, stomach discomfort, anemia, nervous system effects
Disinfection by-products	Compounds formed due to side reactions during the disinfection process	Process of disinfection	Total Trihalomethanes (TTHMs), Haloacetic acids	Liver and kidney problems, increased risk of some cancers

NOTE: For a complete list of contaminants, go to https://www.epa.gov/ground-water-and-drinking-water/national-primary-drinking-water-regulations.

health impacts (it is informative to review the complete list of contaminants on the EPA's NPDWR site).[21] Human activity from industry, agriculture, and wastewater are the dominant sources. Health impacts vary, but many contaminants are responsible for chronic, long-term diseases, including higher risks of cancers and organ damage, and neurological and developmental impacts. The NPDWR inclusion of 90+ contaminants highlights the universal solvent property of water and the ease with which it gets contaminated. While this is a long list, the NPDWR list does not yet include many chemicals detected in drinking water sources and for which there is growing evidence of human and ecosystem health impacts. Much evidence argues that this list should be even longer to include these chemical threats to drinking water quality and public health, but why these chemicals are not on the list will be discussed later in this and other chapters.

Two of the categories of contaminants listed in table 5—disinfectants and disinfection by-products—are ironically a result of drinking water treatment protocols intended to protect the consumer from pathogenic contamination. As discussed in chapter 4, in the early 20th century, the introduction of disinfection by chlorination had significant positive effects on public health and lowered mortality rates from waterborne infectious diseases. However, since those early days, studies have demonstrated that attention must be paid to the process of disinfection. Optimizing the level of added chemicals to ensure effective disinfection through chlorination requires careful attention to specific water quality parameters.[22,23] The drinking water that leaves the treatment facility should have some chlorine (termed residual chlorine) to minimize the chance of recontamination by pathogens as the water is distributed through the pipes to end users. As a result of this residual level, you may get a "chlorine" odor when you drink water right out of the tap. While this residual chlorine is important to protect the consumer from pathogens, maintaining this level can be tricky as too much chlorine can also be harmful. In addition, reactions between the disinfection chemicals and compounds that may be present in the water can form "disinfection by-products," which are known to have health impacts (see table 5). The benefits and challenges of chlorination disinfection highlight the delicate balance of treatment protocols that maximize health benefits and minimize potential side reactions that may have negative impacts. Some drinking water systems have recently switched to other disinfection methods, such as ultraviolet radiation and ozone. These treatments lower the chance of harmful disinfection by-products forming. However, these are more expensive methods of disinfection and highlight the complexity and the cost of compliance with the mandatory standards.

For each primary contaminant, the regulations dictate a maximum contaminant level (MCL), which is "the highest level of a contaminant that is allowed in drinking water."[24] The standards also include a maximum contaminant level goal (MCLG), which is "the level of a contaminant in drinking water below which there is no known or expected health risk. MCLGs allow for a margin of safety and are non-enforceable public health goals." Further, "MCLs are set as close to MCLGs as feasible using the best available treatment technology and taking cost into consideration."[25] The chemical dioxin, introduced in chapter 3, has an MCL of 0.00000003 mg/L (milligrams per liter) but an MCLG of zero. An MCLG of zero means that, ideally, a person should not be exposed to any dioxin through their drinking water to guarantee no harm from this chemical. However, setting the MCL of dioxin at 0.00000003 mg/L is a balance between the practicalities and costs of ensuring that no dioxin is present and protecting the majority of people who may over their lifetime consume water that contains less than 0.00000003 mg/L of dioxin without adverse outcomes.

Setting an MCL, then, is a balance between protecting the health of the population, ensuring that there are no technical challenges in meeting the MCL, and a cost-benefit analysis that compares the cost of meeting the standard to the benefits gained in protecting the health of the population. There may be a small percentage of a community's population—for example, young children, the elderly, or people who are immunocompromised—whom the MCL may not adequately protect. The MCLG protects this "sensitive" population.[26] Regulations that dictate allowed lead levels in drinking water highlight the tension of protection versus the cost of compliance. Children are particularly susceptible to the effects of lead poisoning, and studies indicate that there is no safe level of lead for this population (chapter 6 discusses lead contamination in drinking water). In the United States, drinking water systems are required to manage water parameters such that levels of lead in drinking water are below 15 mcg/L (15 ppb [parts per billion]). For lead, this is referred to as the "action level."[27] Yet, for a child, this level is above what they should be exposed to (the MCLG for lead in drinking water is zero). The World Health Organization (WHO) recommends a maximum concentration of 10 ppb for lead in drinking water. This recommendation balances health protection and feasibility in meeting a standard.[28] The European Union adopted the WHO standard of 10 ppb, and in 2029 will strengthen this standard to 5 ppb.[29,30] In the United States, as dictated by the definition of the MCL, adopting a more stringent standard of 10 or 5 ppb for lead will require the EPA to carry out a cost-benefit analysis.

THE PATH FROM SOURCE TO TAP

The path from source to tap starts with identifying a freshwater source. If this freshwater source is surface water such as a river or lake, this is where the CWA intersects with drinking water quality. The CWA aims to protect surface water by dictating the quality of water discharged from sources, including wastewater treatment plants, industries, and other "point sources." The EPA defines a point source as a "discernible, confined and discrete conveyance, including but not limited to any pipe, ditch, channel, tunnel, conduit, well, discrete fissure, container, rolling stock, concentrated animal feeding operation, or vessel or other floating craft, from which pollutants are or may be discharged. This term does not include agricultural storm-water discharges and return flows from irrigated agriculture."[31] While agricultural discharges may not be considered to emerge from a "single point" on a farm, as we will see in chapter 6, this is a weakness of the CWA and has consequences for drinking water quality. Another weakness is that the CWA often does not protect groundwater sources.

All freshwater sources, whether surface or groundwater, designated for drinking water use are also protected by the Source Water Protection Program under the SDWA.[32] As a result of the CWA and SDWA, ideally, wastewater, industrial and agricultural effluents, and runoff from streets should minimally impact the quality of freshwater sources designated for drinking water. It should be noted that the term *ideally* is intentionally used here, as this is often not the case, as discussed later in this and the following chapters.

Once a freshwater source is designated as a drinking water source, the next step is to assess this water source's physical, chemical, and biological parameters. These data inform the design of the treatment plant such that the water that leaves the plant meets the federally mandated drinking water standards discussed above. While the specifics of a drinking water treatment plant vary depending on the measured physical, chemical, and biological parameters of the water source, typical steps in treating drinking water are discussed below.[33,34,35,36] Figure 12 shows a schematic of the typical steps involved in drinking water treatment.

i) *Screening and Settling:* Water from the freshwater source is screened to remove large debris. This water is then collected in tanks to slow down the flow and allow for some settling and sedimentation of soil and other particles present in the water.

ii) *Coagulation/Flocculation/Sedimentation:* In this step, suspended particles such as soil, which are too small to naturally sediment, are treated with chemicals that aid the sedimentation process. Soil particles tend to be negatively charged. Chemicals, called coagulants, help neutralize the negative charge and allow these suspended particles to aggregate. These coagulants include ionic compounds such as aluminum sulfate or ferric chloride. The positively charged aluminum or iron ions, once dissolved in water, bind with the negatively charged soil particles, neutralizing the charge. These particles now coagulate to form larger and heavier particles, called floc, settling to the bottom of the tanks. These steps lower the turbidity, a measure of the suspended particles in water, and levels of microorganisms in the water.

iii) *Lowering hardness:* In some locations, the freshwater source's local geology may make the water "hard." Hard water measures the levels of calcium and magnesium ions in the water. You may experience this "hardness" as a "mineral" taste from your water. When you use soap to wash your hands, it may feel like a film of residue is left behind as you rinse your hands. Ions, such as calcium and magnesium in the water, react with the molecules that make up soap. This reaction precipitates a chemical that we refer to colloquially as "soap scum," which is harmless but perhaps not a pleasant feel on our hands. While neither calcium nor magnesium ions are toxic, high levels can affect the water's taste and clog pipes. If the water is deemed hard and could potentially impact infrastructure, drinking water utilities may use chemicals to lower the levels of calcium and magnesium

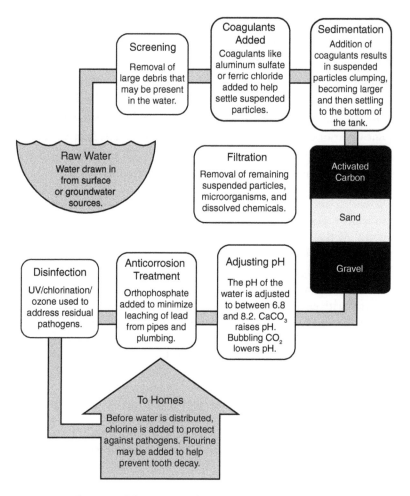

Figure 12. Schematic of the steps involved in a drinking water treatment facility, using water from surface or groundwater sources.

Illustration by Ryann Abunuwara

ions. A common method is liming, where lime, or calcium oxide, is added to the water.[37] Calcium oxide raises the pH of the water to a level considered basic.[38] At basic pH levels, the calcium and magnesium ions precipitate out as insoluble compounds, lowering their dissolved levels and "softening" the water.

iv) *Filtration:* This includes different filtration stages as dictated by the chemical and biological parameters of the water. Gravel and sand filters remove suspended particles that did not coagulate and sediment in the previous steps. The sand filter also removes some microorganisms. Dissolved chemicals require additional filtration. Organic compounds can be removed by granulated activated carbon, which has been treated to have molecular-size pores. Many organic compounds are preferentially attracted to the surface of these molecular-size pores and are filtered out as the water flows through the activated carbon. One process used to remove toxic ions such as cadmium, nitrate, lead, and chromium is ion exchange. An ion exchange filter comprises a solid matrix that contains ions that are nontoxic to humans (such as sodium, potassium, or chloride ions). The ions cadmium, lead, nitrate, and chromium preferentially attach to the filter medium by displacing the nontoxic ions. Filtration methods such as reverse osmosis and nanofiltration are less likely to be used in conventional drinking water systems as these methods result in high water throughput loss and are energy intensive (see chapter 9 for more details).

v) *Adjusting pH:* Utilities closely monitor and maintain the treated water's pH between 6.8 and 8.2 to lower the risk of contamination from ions such as lead and copper and avoid aesthetic effects. Acidic pH levels can cause the leaching of metals such as iron, copper, and lead into the water. Leaching refers to the process by which the metal is converted into an ion, for example, metallic lead converted to lead ions, which dissolve in water. Since metal ions like lead and copper are toxic, the pH of the water must be regulated to levels that minimize the risk of leaching. Some drinking water treatment steps also influence the pH of the water. For example, water treatments such as liming, which reduces hardness, make the pH basic, causing aesthetic effects (e.g., taste, feel). By monitoring the pH of the water, utilities assess what measures are needed to adjust the pH. If the water is acidic, the pH can be raised by flowing the water through a filter containing minerals such as calcium carbonate ($CaCO_3$). If the pH is basic, carbon dioxide (CO_2) may be bubbled through the water, lowering the pH.

vi) *Anticorrosion treatment:* This is a crucial step, particularly in towns with lead-based distribution pipes and homes that may have lead-based plumbing fixtures. While controlling pH is one way to prevent lead from leaching into the drinking water, adding a compound called orthophosphate helps lower the risk of lead leaching from pipes into the drinking water. In places that have lead in the distribution system, the orthophosphate reacts with the lead to form an insoluble film of lead orthophosphate on the insides of pipes and plumbing. The film minimizes the contact between the water flowing through and the lead in the pipes and fixtures and lowers the risk of lead ions leaching into the water.

vii) *Disinfection:* This is the last step in the treatment process, right before the water is released into the distribution system to end users. Methods of disinfection include chlorination, ozone, and ultraviolet radiation. Disinfection kills any residual pathogens not addressed in a previous step. UV disinfection, while more expensive, is a preferred choice as some pathogens such as *Cryptosporidium* and *Giardia* are resistant to chlorination. As the water leaves the treatment facility, the addition of chlorine helps prevent the growth of microorganisms in the interior of the pipes. As a result, the water you drink from the tap may have a faint smell of chlorine.

The treated water is tested before being sent to distribution pipes to ensure that all primary contaminants' levels are below their MCLs. EPA-approved analytical methods are used for these tests. The EPA must certify the laboratory conducting these tests as having the technical expertise and equipment to perform these analyses.

Water samples are collected at the end of the distribution system before entering the end user's plumbing and analyzed to verify that the quality of the water is still in compliance with EPA standards. It is essential to recognize that public water systems (PWS) are responsible for ensuring that drinking water standards are met until the point of connection with end users' (homes, businesses, or industries) systems.[39] Any change in water quality resulting from private plumbing is not the responsibility of the PWS. For example, older buildings and homes may have lead-based plumbing or fixtures. As it passes through these pipes and fixtures, the water could potentially cause lead to leach out and contaminate the water even if the water delivered to this location meets standards. Once again, this highlights the ease with which water gets contaminated, even posttreatment.[40]

The schematic in figure 13 demonstrates that almost every sector of society has a role in drinking water management. The central goal is to deliver safe, reliable drinking water to communities. Water from a protected freshwater source is treated to meet physical, chemical, and biological standards dictated by the NPDWR. Once used by end users, this water becomes wastewater. The CWA regulates the treatment and quality of the discharge of this

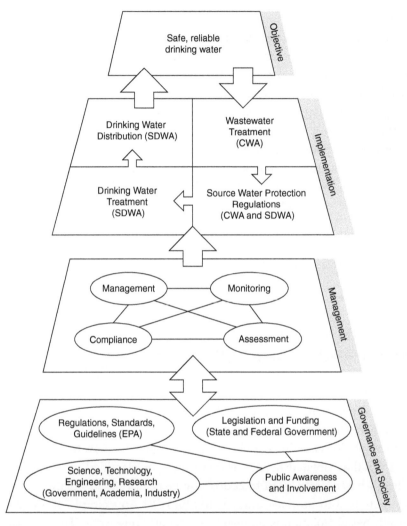

Figure 13. Schematic of the processes and sectors of society that play a role in ensuring the delivery of safe and reliable drinking water to a community.

Illustration by Ryann Abunuwara

wastewater. Local or municipal water utilities are held accountable for managing, monitoring, and assessing drinking water treatment facilities to comply with regulations. State officials are responsible for monitoring local municipal utilities. In turn, the EPA holds states accountable for ensuring that local utilities comply with mandatory regulations. The EPA shapes and informs its regulations, standards, and guidelines by drawing from academic, industry, and government research. Public involvement also plays a crucial role in shaping policies as the EPA is required to receive feedback and comments from the public on its policies, address public concerns, and communicate with the public. Finally, local, state, and federal governments dictate the level of funding allocated to drinking water management.

SUCCESSES OF THE SDWA

Before the passing of the SDWA, drinking water quality was widely variable across the United States. A 1969 Public Health Survey revealed that only 60 percent of public water systems (PWS) delivered drinking water that met federal guidelines.[41] A 1970 report revealed that 90 percent of PWS did not meet guidelines for microbial contamination and studies from the 1960s and '70s raised concerns about the presence of chemical contaminants in drinking water.[42]

As a result of the SDWA and amendments, the number of drinking water violations has decreased. The data in figure 14 show the percentage of PWS that violated health-related standards classified as "acute" and "health-based" from 2011 to 2020.[43] Acute health-based violations "have the potential to produce immediate illness."[44] An example of an acute health-based violation is pathogens in the drinking water. Health-based violations include nonachievement of MCLs or failure to implement treatment protocols that "specify required processes intended to reduce the amounts of contaminants in drinking water."[45] A health-based violation does not necessarily mean that people's health is in jeopardy. For example, a short-term violation of an MCL for a chemical contaminant may not necessarily be of immediate concern, as typically, health consequences result from long-term exposure to chemicals. But such a violation does require immediate action and response by the PWS.

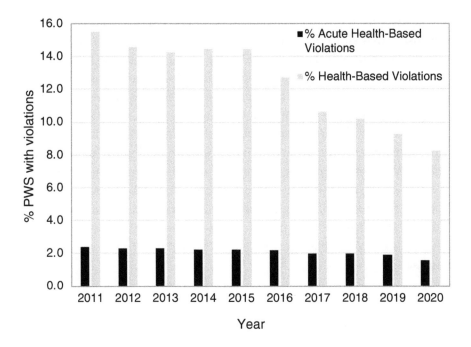

Figure 14. Data showing the percentage of public water systems (PWS) in the United States with acute health-based and health-based violations from 2011 to 2020. The percentage of PWS reporting these violations has decreased over these years.

SOURCE: US Environmental Protection Agency, "Analyze Trends: EPA/State Drinking Water Dashboard," https://echo.epa.gov/trends/comparative-maps-dashboards/drinking-water-dashboard?view=activity&state=National&yearview=FY&criteria=adv&pwstype=Community%20Water%20System&watersrc=All.

On a positive note, over these 10 years, health-based violations have declined. In 2020, 90 percent of people received water from a PWS which met the primary drinking water standards. Since the 1960s, when only 60 percent of the utilities met guidelines (which were not as stringent as current standards), and 90 percent of the water exceeded microbial guidelines, these data demonstrate that drinking water quality has improved. The passing of the SDWA has provided safe drinking water to the people residing in the United States, protecting public health and supporting social and economic development.

At the same time, the data in figure 14 reveal that more work remains to ensure that all PWS meet the mandatory standards of the SDWA. Of particular concern is the ability of PWS to conduct regular and consistent monitoring of drinking water quality and produce timely reports to state agencies and the public. Lowering the incidence rates of health-related violations, improving the ability of PWS to carry out the necessary checks and balances to ensure the safety of the water, and reporting information to the public in a timely fashion requires more funding be allocated to these PWS. Better oversight and support of PWS by local, state, and federal levels of government are also necessary. Ultimately, improving compliance requires an informed public to hold their elected officials accountable for the quality of their tap water and advocate for increased funding to enable all PWS to manage drinking water treatment facilities effectively and safely.

FAILURES IN ACCESS AND THREATS TO SAFE WATER

While the majority of people living in the United States have safe water, inequities prevail in who does or does not enjoy this access.[46,47,48,49,50] As discussed in the previous chapter, access to safe drinking water supports a community's health, educational, social, and economic well-being. Ensuring that these inequities in access to safe water are addressed must be a priority in the United States. Failures in the current regulatory system and new threats could have a widespread impact on drinking water quality. As a result, those who currently benefit from and likely take this access for granted risk losing this privilege. Urgent action by policy makers at the local, state, and federal levels is needed to address these issues so that all residents enjoy the benefits made possible from access to safe water.

Inequities in Access to Safe Water

The fact that in 2019 about 10 percent of PWS violated health-based regulations is concerning. The process from source to tap is complex, demands expertise, and is expensive. A 2018 study analyzing data of health-based drinking water violations in the United States over about 30 years concluded the following:[51]

- There are geographic "hot spots" across the United States, particularly the Southwest, with higher incidences of violations.
- Some water systems struggle with repeat violations.
- The socioeconomics of a community correlates with compliance; low-income, rural areas often have more violations than higher income, urban areas.
- Minority, low-income populations have higher rates of violation of total coliform levels, an indicator of microorganism contamination.

As the data in figure 15 demonstrate, rural and low-socioeconomic communities struggle the most in meeting drinking water regulations.[52] These communities tend to be smaller in population than urban or suburban communities. Revenue generated through water rates is the primary funding source for utilities to maintain and upgrade drinking water systems, conduct the required assessments, and report to state agencies and the public.[53,54] The low number of people served by smaller systems makes it hard to recover the costs associated with drinking water delivery.[55] New and evolving regulations make it all the more challenging for small systems to meet compliance. The data in figure 15 make this evident, as shown by the rise in violations in specific years following the introduction of new regulations. As drinking water systems adapt to these new regulations, incidences of violations increase, with the highest increase in rural, low-income communities.[56,57] Financially strapped utilities are unable to comply with mandatory regulations. The consequence of insufficient funding and lack of expertise is a higher level of drinking water violations that may impact a community's health.

Public Health versus the Cost of Compliance

The definition of the maximum contaminant level (MCL) reveals a tension between protecting public health and the cost incurred in meeting the mandatory standard. The EPA compares the cost of harm incurred from exposure to a contaminant to the cost of compliance. This cost-benefit

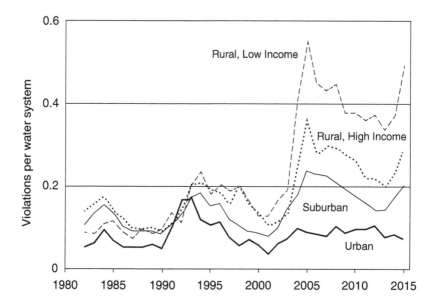

Figure 15. Violations of federally mandated regulations by public water systems in the United States between 1982 and 2015. The highest number of violations are in systems located in rural, low-income communities. As discussed in the text, the spikes in violations following specific years (1990, 2002, 2012) coincide with new regulations and utilities adapting to meet compliance.

Reproduced with permission from M. Allaire, H. Wu, and U. Lall, "National Trends in Drinking Water Quality Violations," *Proceedings of the National Academy of Sciences* 115 (9): 2078–83.

analysis aims to protect the majority of the population. However, some communities, such as the elderly, immunocompromised, and infants, may not be adequately protected by the MCL, as illustrated above with the example of lead in drinking water. The MCLG protects more people than the MCL, but the EPA assesses whether meeting the MCLG would be cost-prohibitive. As a result, MCLs may be set higher than MCLGs.

Increasing Use of Synthetic Chemicals Contaminants

Despite the concerns raised in the 1960s about the increasing number of synthetic chemicals found in water sources, the chemical industry continues to churn out chemicals. Chemicals leach from landfills into groundwater and surface water. Unregulated contaminants from sewage and industrial effluents have been detected in water sources. Chemicals are introduced in products such as pesticides, personal care products, and pharmaceuticals before the impacts of the chemicals on ecosystems and humans can be understood, evaluated, and assessed. Chemical contaminants in water sources and drinking water are a rising threat to public and ecosystem health. Scientific studies point to concerning synergistic effects of mixtures of chemicals on ecosystem and human health. Here synergistic effects refer to some chemicals altering the toxicity of other chemicals when present together, for example, in a water source.[58,59,60] These mixtures are "accidental" in that chemicals released by different sources may all end up in a body of water and potentially trigger synergistic effects. Our current approach of regulating individual chemicals with MCLs may be insufficient, and drinking water regulations may have to be updated to account for mixtures.[61] The challenge is that the science of assessing these synergistic impacts is complex, and understanding how to reframe drinking water regulations with mixtures in mind are ongoing discussions.[62] Chapter 7 will look at the precautionary principle and how such an approach may help address this challenge.

Regulatory Challenges

A 1996 amendment to the SDWA established the Contaminant Candidate List (CCL).[63] The CCL is "a list of contaminants that are currently not subject to any proposed or promulgated national primary drinking water regulations, but are known or anticipated to occur in public water systems."[64] The SDWA requires the "EPA to make regulatory determinations for at least five contaminants from the most recent CCL within five years after the completion of the previous round of regulatory determinations."[65] The purpose of this list is to establish a process for evaluating whether contaminants identified in drinking water sources are health risks and whether or not regulatory action is necessary. Since the passing of the 1996 amendment, there have been five CCLs. CCL 1, established in 1998, included 10 microbial candidates and 50 chemical candidates.[66] In 2003, after the required five years, the EPA concluded that sufficient information existed on 9 of the 60 candidates to make decisions. The EPA decided not to add these 9 to the primary drinking water standards and to continue evaluating the remaining 51. The fact that only 9 out of the 60 candidates were evaluated over five years points to the time needed to assess whether a contaminant should be added to the regulated list. This work is complex and time-intensive. But this also means that while these evaluations are being carried out, people could potentially be exposed to harmful contaminants. Since the 1996 amendment to the SDWA requiring the establishment of CCLs, the EPA has not added any contaminant for regulation as a primary drinking water contaminant.

While the EPA may not have added new contaminants to the primary drinking water standards, the agency has taken some action after incidents of contamination. For example, after a series of microbial contamination cases, the EPA mandated treatment protocols to address microbial contamination.[67] As a result of lead contamination of drinking water in Flint and other cities across the US, the EPA has proposed revisions to the Lead and Copper Rule.[68]

The 1996 SDWA amendment includes the Unregulated Contaminants Monitoring Rule (UCMR). Under this rule, every five years, the EPA issues "a new list of no more than 30 unregulated contaminants to be monitored by public water systems (PWSs)."[69] Public water utilities must add these unregulated contaminants to their monitoring program. While the EPA considers regulatory decisions, studies reveal health risks for some of these compounds. In response to these studies and public concern, the EPA has published health advisories for some contaminants. While these health advisories are not mandatory, the EPA advises public water utilities to heed them and, if necessary, upgrade treatment facilities to address these contaminants.

Climate Change

Climate change is influencing precipitation patterns, causing droughts in many places. In August 2019, the World Resources Institute released a report on the growing challenges that a quarter of the people on the planet will face due to climate change, particularly in severely water-stressed regions.[70] Climate change also impacts water quality. Severe downpours and floods overwhelm existing treatment infrastructure. Sea-level rise and more frequent coastal flooding, both a consequence of climate change, will affect the water quality of drinking water sources. Runoff and floodwater from farms dump pathogens and chemicals like pesticides and fertilizers into drinking water sources. Nitrate and other nutrients used in agriculture end up in waterways and trigger harmful and toxic algal blooms. During unprecedented downpours and floods, sewers exceed capacity, and untreated water is released into rivers and lakes. During recent floods in Michigan, Texas, and Florida, toxic chemicals from refineries, coal ash, industries, waste sites, and agricultural waste were swept by floodwaters into water sources.[71,72] Studies reveal that the burning of homes and businesses from wildfires, increasingly driven by climate change, release toxic chemicals into the air, which rain out and impact water quality.[73] A warmer planet will only add to the other factors affecting water quality, making it more complex and expensive to deliver safe drinking water.[74,75,76]

Investments in Drinking Water Infrastructure

Every four years, the American Society of Civil Engineers (ASCE), a US-based professional society of civil engineers, releases "report cards" on the state of the US infrastructure. In the latest report card, published in 2021, drinking water infrastructure received a "C-minus" grade.[77] This grade is an improvement from reports since 1998 when drinking water infrastructure in the United States received a D grade (in 2005 and 2009, the grade for drinking water infrastructure was D-minus). Since the last report in 2017, there has been more attention paid by local water utilities in upgrading systems and incorporating newer treatment technologies as well as increased federal financing programs. Increased attention and funding for drinking water utilities is a positive indication, and likely a result of the many cases of contaminated drinking water delivered to residents in cities like Flint, Michigan; Newark, New Jersey; and Hoosick Falls, New York.

As part of this assessment, the ASCE also makes specific recommendations for Congress and agencies charged with drinking water management, including the following:[78]

- Triple the annual appropriations to federally funded drinking water programs.
- Integrate new technologies and real-time sensors to increase safety and resilience of drinking water treatment utilities.
- Increase federal and local funds and support training for a new generation of drinking water utility workforce.
- Conduct economic analyses to determine appropriate rate revenues that reflect the "true cost of water that is needed to provide safe, reliable drinking water and more resilient infrastructure."
- Expand funding for affordability programs so that low-income and vulnerable communities do not disproportionately bear the rising costs of drinking water as utilities upgrade and expand drinking water treatment infrastructure.

While a C-minus is an improvement in grade, clearly this is insufficient given the crucial role of safe drinking water in safeguarding public health and well-being. In the United States, the construction of much of the drinking water infrastructure occurred 75 to 100 years ago when cities first set up treatment facilities and distribution systems to address the spread of waterborne diseases. While this infrastructure has served the country well and provides safe drinking water to most people in the United States, this will no longer continue to be the case without serious investments to address failing infrastructure. In November 2021, President Joe Biden signed a $1.2 trillion infrastructure bill. This bill includes the funding of $110 billion for water infrastructure, which provides for the replacement of lead service lines and increased resiliency of water systems from droughts and floods. These funding allocations are important, but much more investments are needed. A 2021 study by the ASCE reported a funding gap of over $1 trillion for water and wastewater treatment systems in the United States between 2020 and 2029.[79] This study indicates that improving the grade for drinking water infrastructure in the United States will require significantly more investments.

The drinking water management system in the United States is by no means perfect. The government must strengthen the current regulatory system. For example, since 1996, no contaminant has been added to the primary drinking water standards list despite mounting evidence of the health impacts of certain chemicals detected in drinking water sources.

Persistent inequities in access to safe drinking water demand that every level of government pay serious attention to ensure that everyone living in the United States, regardless of their zip code, has safe, reliable drinking water. The ASCE report strongly advises that urgent attention must be paid to replace and upgrade the failing infrastructure. Some policy experts have suggested that the approaches established in the 20th century may no longer serve the purpose for 21st-century challenges to drinking water and that there is an urgent need to innovate drinking water management.[80]

Despite these challenges, it is also essential to reflect on the successes of the SDWA. The positive impact of the passing of the SDWA on public health, which in turn advanced the nation's social and economic development, cannot be understated. Recognizing the accomplishments of the SDWA is critical to ensuring that we, the public, are aware of the investments that were made to reach this level of safe water quality for the majority of residents. We do not want to revert to times when our health was compromised due to unsafe water. It is also incumbent on those who have benefited from this privilege to advocate that this access be equitably enjoyed by all. According to the US Water Alliance, "Water equity occurs when all communities have access to safe, clean, affordable drinking water and wastewater services; are resilient in the face of floods, drought, and other climate risks; have a role in decision-making processes related to water management in their communities; and share in the economic, social, and environmental benefits of water systems."[81]

Too often, we forget past challenges. Many of us residing in countries where we have access to safe drinking water take this for granted. Doing so is dangerous as we then do not insist that our political representatives pay attention to crumbling infrastructure or declining funding, particularly when other challenges seem more pressing. Absent safe drinking water, we will not have the health or the social and economic capacity to address these other challenges. The next chapter presents three case studies of contaminants that have challenged and continue to challenge our access to safe drinking water. These cases are presented as examples to learn from—how did these contaminants get into drinking water, what were the health, social and economic costs, and how can we learn from these recent events to develop more innovative, evidence-based policies and water management strategies.

6   Learning from Drinking Water Contamination Events

Chemistry is about interactions and reactions between molecules. An example is the delicate balance of hydrophilic and hydrophobic interactions between phospholipid and water molecules, discussed in chapter 2, which establishes the structural integrity of a cell yet maintains an aqueous environment internally and externally. However, this aqueous environment so necessary for life is also potentially a liability. Chemicals may dissolve in water at levels that exceed their toxicity limit, potentially causing cellular damage. Because water is the universal solvent, the list of regulated chemicals on the primary drinking water standards is long. This list should be even longer as an increasing number of unregulated chemicals are detected in drinking water sources, and studies implicate many as potentially harmful to humans.[1]

This chapter presents three cases of chemical contamination of drinking water. First, we will look at lead in drinking water, a historic challenge since the Roman Empire. The second case is nitrate, used in synthetic fertilizers, a significant threat worldwide and harmful to humans and ecosystems. Finally, the third case is an example of a situation in which chemistry intended to improve our quality of life is now potentially harming us as a growing number of people are exposed to a class of chemicals called per- and polyfluoroalkyl substances (PFAS) through their drinking water. These cases highlight the ease with which water gets contaminated and how historical legacies, industry interest, and profits influence drinking water quality with consequences for public health.

Here in the United States, unsafe levels of lead were present in the drinking water delivered to residents of cities such as Washington, DC; Flint, Michigan; and Newark, New Jersey. Communities surrounded by farming-intensive regions in states such as Iowa, Kansas, Oklahoma, and California have unsafe levels of nitrate in drinking water sources. Hoosick Falls, New York; Fountain, Colorado; Wilmington, North Carolina; and Parkersburg, West Virginia, are examples of towns where PFAS have been detected in the drinking water due to local industries using or manufacturing these compounds. The intent behind discussing these cases is not to alarm but rather to raise vigilance and inform how and why we need to strengthen regulations on the use of chemicals to protect drinking water. We must learn what went wrong in such cases and we, the public, who consume this water, and whose health is at risk, must understand the critical role of strong, evidence-based regulations and actions.

These cases also highlight the importance of learning from "unintended consequences" as a result of past practices. Arguably, we must recognize that the term *unintended consequence* is increasingly obsolete as our knowledge and understanding have advanced. Further, decisions will always have consequences, and actions and policies must be more deliberate in accounting for and minimizing potential adverse outcomes.

"LIQUID SILVER"—PLUMBUM (PB)

Lead is an abundant metal on Earth. It is a relatively soft, malleable metal with a low melting point. The Latin name for lead is *plumbum*, which translates to "liquid silver," as it coexists with silver in the ore galena. It is also a neurotoxin and can impact brain development in infants and children, resulting in behavioral challenges and lower IQ. Exposure to lead can stunt growth and cause hypertension, hearing loss, infertility, anemia, digestive issues, decreased life span, and, at high enough doses, death in children. Lead exposure in a pregnant woman can trigger mutations in the fetus's DNA, which the grandchild may inherit.[2] While an understanding of why lead is toxic is relatively recent, the ancient Greeks were aware that lead might be harmful.[3] Yet, lead's properties, which make it a strong, yet malleable material took precedence over its toxicity.

Since ancient times, coins, vessels, cosmetics, jewelry, construction, piping, weaponry, ceramic glazes, paints, and pigments had lead. The Romans initiated large-scale use of lead in water pipes. Before lead, terracotta or wood was used to make water pipes. Neither of these materials is sturdy. Lead appeared to be the ideal material as lead pipes are durable. Lead's low melting point and malleability allow it to be easily shaped into pipes. The name of the occupation of the people installing pipes, "plumbers," was derived from *plumbum*.

After the fall of the Roman Empire, the use of lead in water pipes declined. However, in the late 19th and early 20th centuries, pipes were once again made from lead. During this time, cities were building filtration and chlorination systems to prevent the spread of pathogens through drinking water. The distribution system had to be robust and leak-proof to prevent contamination. While iron was considered an option, it corrodes when exposed to water and hence, is more likely to leak. Lead does not corrode as easily as iron when exposed to water and thus, seemed a better option for pipes.

Health officials and physicians did raise concerns about using lead in water pipes.[4] But, its many favorable properties overrode the warnings of the medical community. The Lead Industries Association (LIA), established in 1928, promoted and lobbied plumbing associations, waterworks officials, local city boards, and federal agencies to use lead pipes. As a result, many cities adopted codes that required lead piping to deliver treated water.[5]

Engineers and city officials focused on the structural integrity of lead pipes and their durability. The dangers of cholera and typhoid, contracted through contaminated drinking water, were fresh in people's minds. Durable pipes would protect from contamination by pathogens. Officials were not as concerned about the toxicity of lead as it was not immediately apparent that people were getting sick from drinking water when using lead pipes. A 1938 article in the *Journal of the American Water Works Association* highlighted skepticism to the dangers of lead: "If the very small amounts [of lead] which persons ingest by drinking water and eating food, were as harmful as some people believe them to be, there would be many more cases of lead poisoning that are known to occur."[6] Such statements highlight the challenges of chemical contamination, the effects of which are often slow and symptoms not apparent for years. As we now know, there is no safe level of lead. Studies reveal that very low levels of lead in the blood can cause behavioral and learning difficulties in children.[7]

With growing concerns about lead over the 1930s and 1940s, some cities slowly moved away from installing lead pipes. However, until the 1980s, building codes continued to recommend using lead for water pipes. A 1984 survey conducted by the EPA of 153 public water utilities, many serving over 100,000 residents, revealed that 73 percent had lead pipes. It was not until 1986 that an amendment to the Safe Drinking Water Act (SDWA) banned lead in water pipes, solder, and plumbing fixtures.[8,9] In 1970, the United Kingdom banned the use of lead pipes but did not prohibit lead solder for connections until 1987.[10] In Canada, the National Plumbing Code banned lead pipes in 1975 and lead solder in 1986.[11]

So, if the SDWA banned lead pipes in 1986, why is lead still contaminating drinking water? Why were residents of towns like Flint, Newark, and Washington, DC, exposed to lead in their drinking water? Many of the lead pipes laid down in the 1900s to distribute water within a city (also called lead service lines) remain. In the United States, nearly a third of municipal water systems continue to rely on lead service lines.[12] Residences, schools, and businesses built before the 1986 ban on lead may still have internal lead pipes, plumbing fixtures (such as brass, which contains lead), or lead solder in their pipes. The only way to definitively eliminate this risk is to replace all lead service lines and lead-based pipes and fixtures. In the meantime, preventing lead from leaching into the water requires strict adherence to drinking water treatment protocols. In the United States, the 1991 amendments to the SDWA included the Lead and Copper Rule, which requires municipal water treatments to abide by treatment protocols, also known as corrosion control, that minimize leaching of lead from pipes and fixtures into water.[13] The primary corrosion control treatments are regulating the pH of the water and the addition of chemicals, both of which lower the risk of lead ions leaching into the water.

The pH of a solution influences the solubility of lead. Pure water has a neutral pH of 7. At this pH, the solubility of lead in water is low. However, the pH of water can change because of some dissolved chemicals. Acidic solutions, which have a pH less than 7, increase the solubility of lead in water. In an acidic solution, the H^+ ions can react with the lead atoms in the metal to form lead ions; these lead ions are soluble in water. Water from rivers and lakes may be acidic, and some water treatment steps may also acidify the water. The pH of treated water is carefully monitored and controlled to levels that lower the risk of lead dissolving.

Ions, such as chloride, which may be present in some water sources, also influence lead's solubility. Like the hydrogen ion, chloride ions react with lead metal and form soluble lead ions. Utilities use a chemical called orthophosphate to minimize reactions that may cause the leaching of lead into water. The orthophosphate reacts with the lead in pipes and fixtures, forming an insoluble, protective film along the inner walls. This film prevents dissolved compounds in water like acids or chloride ions from reacting with the lead and leaching into the water.

Despite the Lead and Copper Rule, communities in cities such as Washington, DC, Flint, and Newark have been exposed to unsafe levels of lead in their water. The fact that these cities have a significant Black population is essential to recognize.[14]

In 2004, the *Washington Post* broke the news to DC residents that their drinking water might contain unsafe levels of lead.[15] According to this newspaper, the DC Water and Sewer Authority had been aware of a problem since 2002, which meant that some residents might have been consuming water that contained unsafe levels of lead for over two years. The DC system relied on regulating the pH of the water to minimize corrosion from lead pipes and fixtures. This approach seemed to keep levels of lead in drinking water in compliance. In 2000 the utility decided to change the chemical used to disinfect water. Chlorine, which was used in DC to disinfect the water, can react with dissolved organic compounds, forming disinfection by-products. Some of these by-products are a health risk and are regulated contaminants. The utility switched to chloramine, which lowers the risk of the formation of disinfection by-products. Unfortunately, chloramine reacted with the lead in the pipes, leaching lead ions into the water and increasing lead levels in the city's drinking water.[16] In 2004, the utility began using orthophosphate as the corrosion control treatment protocol. Subsequently, lead levels in the water decreased and met compliance levels.

City officials in Flint, Michigan, a town going through bankruptcy and under emergency management, made a series of erroneous and negligent decisions about managing its drinking water system.[17,18] To save money, the emergency manager decided to switch the drinking water source from Detroit's water utility, which had been supplying Flint with drinking water. The city upgraded a local drinking water treatment facility and selected the Flint River as the drinking water source. In April 2014, the city began receiving treated water from the Flint River. Almost immediately, residents complained about the water's color and odor and some people experienced skin rashes. City officials ignored residents' concerns and insisted the water was safe. Within a couple of months of the switch, the water utility notified residents that *E. coli* might be present in their water and that they should boil the water. In January 2015, the water utility detected unsafe levels of trihalomethanes (THMs), a by-product of chlorine disinfection.

As early as February 2015, tests revealed that tap water samples from one home had high levels of lead. Yet, officials did not take this as a widespread issue. Residents, increasingly concerned about their tap water and the lack of response from city officials, reached out to a research group at Virginia Tech known for their work on drinking water contaminants such as lead. In September 2015, working in collaboration with residents from Flint, the Virginia Tech research group released data revealing alarming levels of lead in tap water samples collected from residences in Flint.[19] That same month, a local pediatrician raised concerns that blood lead levels in children in the community had increased since the drinking water switch.[20] Initial responses by the city and state officials were to deny the validity of these results. In October 2015, after state officials confirmed these results, the city was ordered to switch back to the Detroit water system.

A "cost-saving" decision to switch the drinking water source for Flint has resulted in state and federal allocations of hundreds of millions of dollars to provide bottled water, filters, and replacement of lead service lines and plumbing. This financial cost does not account for the health and social costs borne by Flint residents and the potential long-term impacts on the children exposed to unsafe levels of lead. A particularly egregious decision by the Flint water utility was to not use orthophosphate as part of the treatment protocol to save a few hundred dollars a month.[21,22] This was a dangerous decision as the city has lead service lines. This decision was all the more dangerous as the Flint River's water is particularly corrosive as it has a high level of chloride ions.[23] It appears that the utility did not account for the presence of chloride ions in the Flint River water as part of its treatment protocol. The combined effect of the decision to not use orthophosphate and not addressing the Flint River's chloride levels exacerbated the leaching of lead, resulting in extremely high and dangerous levels of lead in the drinking water. Lead levels as high as 13,200 mcg/L were detected in a resident's home—900 times the regulated level.[24] Understanding the chemistry of a water source is essential in assessing its feasibility as a drinking water source and identifying the treatments necessary to meet drinking water regulations. The people in charge of Flint's drinking water did not have the expertise to make informed decisions about the city's drinking water and prioritized cost-cutting measures over the water's safety. A report released by the Michigan Civil Rights Commission concluded that what happened in Flint was a clear case of environmental and racial injustice. As stated in the executive summary of this report:

> The people of Flint did not enjoy the equal protection of environmental or public health laws, nor did they have a meaningful voice in the decisions leading up to the Flint Water Crisis. Many argue they had no voice.
>
> The Commission believes that we have answered our initial question, "was race a factor in the Flint Water Crisis?" Our answer is an unreserved and undeniable—"yes." We do not base our finding on any particular event. It is based on a plethora of events and policies that so racialized the structure of public policy that it systemically produced racially disparate outcomes adversely affecting a community primarily made up of people of color.[25]

In her book *The Poisoned City*, author Anna Clark draws a direct line between the decades of systemic racism—for example, the practice of "redlining"—experienced by Flint's Black residents and the mismanagement that led to unsafe water.[26] A study by the Legal Defense Fund highlights that "redlining" contributed to lower levels of municipal services in Black communities, including water and sewer systems.[27] Redlining also prevented wealth growth, limiting the financial ability of Black communities to upgrade to lead-free infrastructure and plumbing.[28,29,30]

According to the US Census Bureau in 2019, Flint was 54 percent Black, 39 percent White, and 4.5 percent Hispanic or Latino, with 38.8 percent of residents below the poverty line.[31] Cities such as Flint are increasingly struggling with declining job opportunities, decreasing investments from private and public sectors, a dwindling tax base, and crumbling infrastructure. As discussed in chapter 5, cities such as Flint are increasingly unable to generate the revenues necessary to meet the growing costs for the delivery of safe drinking water. We will revisit these issues and consider possible solutions in chapter 11. Like Flint, residents of Newark were exposed to unsafe levels of lead in their drinking water for well over a year.[32,33,34] The cause was a change in treatment protocols that made the water more acidic. Flint, Newark, and many other communities bear the social and health costs of lead contamination—a legacy of the past lobbying efforts of agencies such as the Lead Industries Association.

A report by the Natural Resources Defense Council (NRDC) revealed that in 2015 over 18 million people received water from drinking water utilities that violated the Lead and Copper Rule.[35] These violations do not necessarily mean that unsafe levels of lead were in the water. Some violations may have been for infrequent testing or failure to make timely reports to the public. These violations reflect how water treatment facilities often fail to carry out the checks and balances necessary to ensure that lead is not present in drinking water, highlighting the complexities inherent in drinking water management and the ease with which water can become unsafe. Ensuring compliance with protocols such as those dictated by the Lead and Copper Rule demands expertise, constant assessment, vigilance, and funding. Failure to do so risks the health of a community.

These lead contamination cases have prompted the EPA to revise guidelines under the Lead and Copper Rule.[36] Yet, water quality experts believe that the EPA's approaches to lead management are simplistic as the guidelines do not account for the multiple ways lead can leach from pipes into drinking water.[37] As scientists and communities advocate, the only way to address lead in drinking water conclusively is to replace all lead service lines and assist homeowners, schools, and businesses in replacing lead-based plumbing. In Flint, as a result of a lawsuit filed against the city of Flint and Michigan, a $97 million settlement was reached to pay for the replacement of lead service lines.[38] Newark has also received funds to replace lead service lines.[39] Some states have embarked on replacement of lead service lines.[40,41] While these state-level efforts are commendable, the federal government must play a role in this so that residents in all states benefit from lead-free water. The Bipartisan Infrastructure Bill, approved by the US Congress and signed by President Joe Biden in November 2021, allocates about $55 billion to drinking water infrastructure, of which $15 billion will be used to replace lead service lines.[42] While EPA estimates suggest that between $28 and $47 billion are needed to replace lead service lines in the United States, these funds allocated by the US Congress are a crucial first step toward lowering the risks of lead exposure through drinking water.[43]

FEEDING THE WORLD

In 1900, the world's population was 1.65 billion; today, it is 7.9 billion. This population growth since the early 20th century is, in no small part, a consequence of the Haber-Bosch process, which produces nitrate-based fertilizers. Fritz Haber was a German scientist who discovered a way to convert atmospheric nitrogen, N_2, into ammonia.[44] The chemical reaction of this process is seemingly straightforward:

$$N_2 + 3H_2 \rightleftharpoons 2NH_3$$

But this is not an easy process. While air contains about 78 percent N_2, the chemical bond between the two nitrogen atoms is strong. As a result, N_2 is chemically stable and does not react easily. For N_2 to react requires extreme conditions of temperature and high pressure. Haber developed a catalytic process that lowers energy demands and speeds up the reaction rate. He also identified experimental conditions that maximize the yield of the reaction between N_2 and H_2 to form ammonia, NH_3. Carl Bosch scaled up Haber's process to feasibly manufacture large quantities of ammonia. Why does this matter?

All life requires a source of nitrogen, a key element of life. While N_2 makes up 78 percent of air, this is typically not a nitrogen source for life.

Most living organisms obtain nitrogen through their food. For humans, our protein sources include meat, fish, poultry, and legumes. For plants, nitrogen sources are minerals in the soil, like potassium nitrate (saltpeter) or waste from animals, like manure and guano (seabird droppings). Nitrogen-fixing bacteria are the one class of living organisms that can convert atmospheric N_2 into ammonia. The root system of legumes is a host for these nitrogen-fixing bacteria, so these plants are a source of nitrogen in the form of proteins.

With the establishment of large-scale agriculture to feed growing populations, it became clear to farmers that maintaining healthy crops and high yields required adding nutrients to the soil. Saltpeter is one such source, as is waste from livestock. Guano is another source. Alternately growing legumes as part of crop rotations in a field helps maintain nitrogen levels in the soil. However, relying on legumes is slow; relying on livestock waste is limiting. Sources of saltpeter and guano are not widely available.

During the late 19th century, the primary sources of natural fertilizers, like saltpeter and guano, were Chile and Peru. Growing concerns about accessing sufficient natural fertilizers triggered scientists to work on ways to produce synthetic fertilizers. The Haber-Bosch method solved this problem by establishing a commercially feasible, large-scale production process to convert atmospheric nitrogen into ammonia. The ammonia is converted to ammonium nitrate and applied as a synthetic fertilizer on farms. Ammonium nitrate serves as the source of nitrogen for crops. Haber and Bosch were awarded the Nobel Prize in Chemistry (in 1918 and 1931, respectively). While practices such as plant breeding, genetic modification, and mechanization have increased agricultural yield, at least 30–50 percent of the increased agricultural productivity since the early 1900s is a result of synthetic nitrate fertilizers produced using the Haber-Bosch process.[45] Estimates suggest that since 1900 about 40 percent of population growth has resulted from synthetic fertilizers.[46,47] The Haber-Bosch method is used even today to manufacture synthetic nitrate fertilizer.

While feeding a growing population is a remarkable feat, the rampant overuse of nitrate fertilizer is now wreaking havoc on ecosystems and is a significant source of water pollution—an example of an unintended consequence. Globally, only about 17 percent of nitrate fertilizer synthesized by the Haber-Bosch process and used in agriculture ends up as nitrogen in the forms of food consumed by humans (crop, dairy, meat).[48] The unused nitrate ends up in runoff from farms and flows into water sources—lakes, rivers, and seas. Since nitrate is a nutrient, this runoff triggers an explosion of plant growth in these bodies of water—algal blooms—which deplete levels of dissolved oxygen. As a result of oxygen depletion, aquatic life suffocates, and large areas of aquatic regions become dead zones.

In addition to damaging aquatic ecosystems, nitrate fertilizers are a concern for drinking water. Ammonium nitrate dissolves in water, forming the nitrate ion, which is toxic to humans. Its toxicity is of particular concern for infants as it can cause methemoglobinemia, or blue baby syndrome. Due to the different pH levels of an infant's stomach compared to an adult's stomach, infants have a different bacterial population. In infants, bacteria can convert nit*rate* to nit*rite*. The nitrite binds to hemoglobin, preventing oxygen uptake, and can suffocate an infant. Studies suggest that exposure to nitrate in pregnant women may cause anemia, spontaneous abortions, and preeclampsia.[49] Some studies also implicate nitrate with higher risks to some cancers.[50] In the United States, the MCL for nitrate in drinking water is 45 mg/L (also expressed as 10 mg-L NO_3-N), the same as the level recommended by the WHO. However, recent studies suggest that there may be increased risks to some cancers at levels lower than this standard for nitrate.[51]

A 2018 report by the United Nations Food and Agriculture Organization titled "More People, More Food, Worse Water? A Global Review of Water Pollution from Agriculture" captures the complex nexus of synthetic nitrate fertilizers and water quality, as indicated in this quote:

> Nitrate in groundwater has been reported as a major problem in Europe, the United States and South and East Asia. In Europe, even when mean concentrations of nitrate in groundwater have remained relatively stable in the last few decades, nitrate drinking water limit values have been exceeded in around one-third of the groundwater bodies for which information is currently available. Additionally, in India, hundreds of districts in 21 Indian states have reported an occurrence of nitrate in groundwater that is well beyond the national permissible limit.[52]

In the United States, drinking water delivered to towns near agricultural areas across many states, including California, Minnesota, Iowa, and

Mississippi, routinely violate the drinking water standard for nitrate. The title of a 2019 news article poignantly captures the burden borne by farming communities—"They Grow the Nation's Food, But They Can't Drink the Water"—which again disproportionately impacts low-socioeconomic communities and communities of color in the agricultural sector.[53,54]

Nitrate can be removed from drinking water through water treatments such as ion exchange and reverse osmosis. These methods are expensive, are energy intensive, require expertise, and generate waste, which needs careful disposal.[55] As a result, such treatments are often out of reach for low-socioeconomic communities. Bioremediation methods that rely on microbial breakdown of nitrate are another solution to lowering nitrate levels in drinking water sources.[56] However, these processes are slow, are complicated to manage, and require additional treatment after the remediation steps.

Feeding the world is a desirable goal. However, given the growing environmental, human health, and economic costs of excessive nitrate use, we must exercise smart and productive farming practices that minimize nitrate use. As highlighted in the title of the news article about farming communities, addressing nitrate overuse is also an issue of social justice—communities who bear the brunt of this overuse are not responsible for nitrate contamination. These are often the communities working on the farms that feed the rest of us or have livelihoods that rely on aquatic ecosystems impacted by nitrate runoff. To protect water sources from nitrate runoff will require regulatory changes and the education of farmers on agricultural methods that minimize or shift away from synthetic nitrate fertilizers. For example, investments in "green infrastructure" solutions can lower the level of nitrate runoff from farms into drinking water sources (see chapter 8). We have the knowledge, understanding, and methods to address nitrate contamination in drinking water.

FOREVER CHEMICALS

In the 1930s, a Dupont chemist, Roy Plunkett, was researching replacements for refrigerant gases as the ones in use at that time were toxic. These replacement chemicals included compounds called chlorofluorocarbons, or CFCs.[57] Plunkett accidentally discovered that a compound used as a starting ingredient for CFCs—tetrafluoroethylene—reacted with itself to form long molecular chains, or polymers. This polymer, polytetrafluoroethylene (PTFE), has intriguing properties. PTFE is slippery, can be used as a lubricant, and has excellent insulation properties. PTFE is highly hydrophobic, so water does not interact with PTFE. Water forms beads on the surface and does not wet PTFE. Soon, many industries, including electronics, aerospace, and communication, began using PTFE. In 1949, Dupont named this compound Teflon and used it to make "nonstick" cooking pans.

At the microscopic scale, a cooking pan's surface has cracks and crevices, and food gets caught in these crevices. This food can burn and stick onto the pan during cooking, making it hard to clean. Oil, used to minimize burning by filling the cracks, is hydrophobic and less likely to interact with water-rich food. As a result, food does not stick to the pan. With growing health concerns about consuming too much oil, materials that made pans "nonstick" were desirable. A minimal amount of oil on the surface of these Teflon-coated pans allows a fried egg to slide off the pan. Chemistry, in this case, using materials that minimized interactions between the surface of the pan made hydrophobic by Teflon and the water molecules in the food, allows for the cooking of a fried egg that does not leave a messy film on the pan and uses less oil.

The discovery of Teflon, with its seemingly amazing properties, triggered research on fluorinated compounds. In 1951, Dupont began using perfluorooctanoic acid (PFOA) to synthesize Teflon. In 1956, 3M recognized the stain-resistant properties of perfluorooctanesulfonic acid (PFOS) and launched Scotchgard stain repellant protection. (Table 6 lists the acronyms and names of the fluorinated compounds in this chapter.) The uses of fluorinated compounds expanded into food packaging. In the 1960s, the US Navy, working with 3M, researched the use of fluorinated compounds in firefighting foams. A tragic fire aboard a naval vessel in which 134 sailors died motivated the Navy to look for more effective firefighting foams. The firefighting properties of foams formed with these fluorinated compounds were deemed superior and became standard use globally in military bases, airports, and firefighting training.

Since the 1960s and 1970s, studies by Dupont and 3M have indicated

Table 6 Acronyms and names of fluorinated compounds discussed in this chapter

Acronym	Name
CFC	chlorofluorocarbon
PTFE	polytetrafluoroethylene (Teflon)
PFOA	perfluorooctanoic acid
PFOS	perfluorooctanesulfonic acid
PFAS	per- and polyfluoroalkyl substances
GenX	ammonium salt of hexafluoropropylene oxide–dimer acid (HFPO-DA) fluoride

that PFOA and PFOS might be toxic to humans. Tests revealed that these compounds were present in the blood of employees at Dupont.[58] There are no natural sources of these compounds, so blood levels had to have resulted from workplace exposure. Studies suggested that PFOA can pass through the umbilical cord to babies in the womb. Animal studies on exposure to these compounds were troubling. Rats, rabbits, and dogs exposed to PFOA had enlarged livers. Yet, none of these results were made public during this time.[59] Animal and human studies suggested higher risks to certain cancers due to exposure to PFOA.[60,61] In 2005, the EPA labeled PFOA as a possible carcinogen. Dupont and 3M stopped manufacturing PFOA in 2005, but Dupont continued to produce replacement PFAS. In 2009, the Stockholm Convention, an international agreement governed by the United Nations Environment Programme (UNEP), restricted the use of compounds called persistent organic pollutants (POPs). POPs are chemicals designed to resist degradation and hence called "forever chemicals." If POPs are released into the environment, they persist for a very long time. In 2017, the Stockholm Convention designated PFOS, PFOA, and related compounds as POPs.[62]

In 2011, the US Department of Defense announced that PFAS contaminated the groundwater at over 500 bases. This contamination was a result of the use of PFAS-based firefighting foams. In 2012, the EPA recommended that drinking water utilities check for the presence of PFAS and placed these compounds on the EPA's unregulated contaminant list to be considered for "future regulatory action to protect public health."[63] In 2016, the EPA set a "lifetime health advisory" of 70 ppt (parts per trillion) of PFOA and PFOS, individually or combined.[64] In 2017, a study was released that indicated that as many as 15 million people in the United States consume drinking water that exceeds this 70 ppt limit.[65]

A study released in 2019 recommended that the combined MCL of PFAS (including PFOA and PFOS) be set at 2 ppt.[66] A 2020 study by the Environmental Working Group revealed that about 18–80 million people in the United States receive tap water containing combined levels of PFOA and PFOS of 10 ppt or greater.[67]

This class of "miracle" fluorinated compounds, PFAS, has exploded beyond PFOA and PFOS. There are now about 4,000 compounds classified as PFAS.[68] PFAS have been detected in groundwater in Australia.[69] Surface and groundwater sources and drinking water in Bangladesh, India, Japan, Indonesia, Vietnam, Nepal, and Thailand have been contaminated by PFAS.[70] A 2019 report revealed that health costs associated with human exposure to PFAS would cost the European Union between €52 and €84 billion per year.[71]

Regulatory response to limit the use of PFAS over the past 70 years since Dupont first began marketing Teflon has been slow. When these chemicals were first used, most countries had not established environmental regulations. However, even after establishing environmental regulations, industry interest superseded public health concerns. The good news is that there have been recent movements toward limiting the use of and regulating these toxic chemicals. In 2015, the Madrid Statement, signed by hundreds of scientists, was a call to "the international community to cooperate in limiting the production and use of PFAS and developing safer nonfluorinated alternatives."[72] In May 2019, the Stockholm Convention agreed to ban PFOS and PFOA.[73] The Convention also advocated a shift toward fluorine-free alternatives that do not persist in the environment and are less toxic to humans and ecosystems. However, despite the evidence of health hazards from exposure to PFOS, some countries argued for five-year extensions for its use in applications like electronics, firefighting foams, and some textiles.

The hesitancy to ban all PFAS and enforce immediate regulations highlights the tension between industry interests and protecting human and ecosystem health. While nonstick coatings on pans limit the amount of oil consumed and fluorinated firefighting foams provide better protection, these benefits must be compared with the health impacts caused by PFAS and products that rely on them. Companies like Dupont that continued to produce chemicals despite data indicating their health impacts is a story that keeps repeating.

In response to the concerns raised about PFOA and PFOS, Chemours, a Dupont spin-off, now manufactures fluorinated substitutes. A replacement for PFOA, GenX, is a salt of the compound hexafluoropropylene oxide–dimer acid (HFPO-DA) fluoride. Chemours has been manufacturing this compound since 2009. The safety of these replacement compounds is under investigation.[74,75] In 2017, a local newspaper reported that GenX had been detected in the Cape Fear River, the local drinking water source for Wilmington, North Carolina, and neighboring towns.[76] An official at the North Carolina Department of Environmental Quality was quoted in the 2017 news article saying, "unfortunately, with these unregulated contaminants, we have one hitting us after another, and we're trying to deal with it." The EPA's response to this situation was that "in its review of the GenX premanufacture submission (for approval to make it), EPA determined that the chemical could be commercialized if there were no releases to water." However, as reported in the local newspaper, this was not the case as GenX contaminated the river.

On the one hand, compared to past practices, there has been more data gathering and monitoring of PFAS in water sources and drinking water, often by the scientific community.[77] Environmental and community organizations are also pushing for action.[78,79,80,81] For example, once newspapers in the Wilmington area published articles about the release of GenX into the local river, officials demanded action by Chemours. However, many questions remain. Why were these chemicals released from the plant when the manufacturing requirement was that they should not be? Why were measures not implemented to monitor the river and drinking water independently as soon as this plant opened? Why are these chemicals still being used when evidence indicates their toxicity at low concentrations?

In January 2020, the US House of Representatives passed a PFAS regulation bill, which requires the EPA to define drinking water regulations for this class of compounds within two years.[82] In February 2020, the EPA released a statement saying that it was "issuing preliminary determinations to regulate PFOA and PFOS," a step in the direction of defining drinking water standards. Further, "the agency is also gathering and evaluating information to determine if regulation is appropriate for other chemicals in the PFAS family."[83] Absent federal regulations, some US states took initiatives to define statewide regulations.[84] New York State proposed a maximum level of 10 ppt for PFOS and PFOA in drinking water. Vermont passed a bill that sets a maximum level of 20 ppt combined for five PFAS.[85] Some states are suing companies to help pay for the costs associated with removing PFAS from drinking water.

In February 2021, the EPA released a press statement that the agency has decided to regulate PFOA and PFOS in drinking water.[86,87] As stated in its announcement of the Fourth Drinking Water Contaminant Candidate List, the "EPA has determined that PFOA and PFOS may have adverse health effects; that PFOA and PFOS occur in public water systems with a frequency and at levels of public health concern."[88] The EPA also recognized that existing analytical techniques have sufficient sensitivities to detect these compounds at levels that pose a risk. Further, treatment methods including granulated activated carbon, ion exchange, reverse osmosis, and nanofiltration effectively lower the level of these compounds in water to within safe limits (some of these methods were discussed in chapter 5; others are discussed in chapter 9). The specifics of the levels at which PFOA and PFOS will be regulated (i.e., the MCLs) have not yet been released, but once these two chemicals are officially on the NPDWR list, these will be the first additions since 1996!

The EPA intends to continue its assessments on whether and how to regulate the broader class of PFAS under the SDWA.[89] In October 2021, the EPA released a plan to tackle PFAS contamination, which recognizes that a "lifecycle approach" is needed, not just addressing the cleanup of PFAS contamination sites but also preventing "new PFAS contamination from entering air, land, and water and exposing communities."[90] And in January 2022, the EPA released a regulation that will require water utilities to monitor for the presence of 29 PFAS in the treated drinking water being delivered to end users.[91] These monitoring data will be used to assess the need for drinking water regulations for this class of compounds. The EPA's actions, while delayed, are much-needed steps to begin the process of addressing PFAS contamination. It is up to the voting public to ensure that the EPA roadmaps and decisions to regulate PFAS translate into real action.

The three cases of contamination of drinking water by lead, nitrate, and PFAS are lessons to learn from and understand what went wrong. These cases demonstrate that the current process of regulating chemicals like lead, nitrate, and PFAS is far too slow, and as a result, the public pays the price. A key lesson then is that we must pay attention to the chemistry of water that makes it so easily contaminated. This chemistry also dictates that treating unsafe water to make it safe is complicated, expensive, and technically challenging. So, are there ways to manage the delivery of safe drinking water that recognize this chemistry? Chapter 7 discusses the precautionary principle's role in drinking water management—a "better safe than sorry" approach. In chapters 8 and 9, we will look at solutions that apply this principle to the delivery of safe drinking water.

7 The Precautionary Principle and Safe Drinking Water

A 2017 article in *Politico* titled "What Broke the Safe Drinking Water Act?" profiled the battle to regulate a chemical called perchlorate and highlighted the complexities and failures of the SDWA.[1] Perchlorate is a chemical that has both natural and industrial sources. This chemical has been detected in water sources in areas near industries that manufacture or use this chemical. Studies indicate that perchlorate can cause damage to fetal and infant brains and can interfere with the functioning of the thyroid gland.[2,3]

In 2011, after 20 years of research and debate, the EPA decided to regulate perchlorate under the SDWA. The defense department and companies with defense contracts challenged this decision in court. As a result, the EPA did not set an MCL. In June 2020, the EPA reversed the 2011 decision and decided not to regulate perchlorate under the SDWA.[4] In its announcement of this decision, the EPA stated that based on "the best available science and the proactive steps that EPA, states and public water systems have taken to reduce perchlorate levels, the agency has determined that perchlorate does not meet the criteria for regulation as a drinking water contaminant under the SDWA." Environmental organizations raised concerns about this decision and filed a lawsuit against the

EPA, countering the agency's decision to not add perchlorate to the list of regulated drinking water contaminants.[5]

Perchlorate and PFAS are just two examples of the 85,000-plus chemicals registered for use in the United States by every sector of the economy—industry, defense, and agriculture—with many ending up in household products.[6] Global estimates of chemicals on the market range from 75,000 to 140,000.[7] Our reliance on such a large number of chemicals may be fine if studies assessing their effects on humans and ecosystems have demonstrated their safety or identified risks. And not all chemicals are of concern. After all, chemicals drive the biochemistry of life.

So, the question then is, how do we know whether the presence of a chemical in drinking water is cause for concern and at what level? While scientists can identify the presence of chemicals in a source for drinking water, screening for the presence of 85,000-plus chemicals in all bodies of water is an insurmountable challenge. Further, identifying a chemical in a water source is not the final answer, but whether it is harmful to humans and ecosystems and at what level.

So, perhaps the focus should be on the other end by assessing the toxicity of chemicals before their use in products—before they end up in a body of water. If research suggests that a chemical may be unsafe, regulatory agencies like the EPA should not approve its use. Suppose a chemical, for which data suggest concerns, is essential for use in a particular process or product, and there are no safer alternatives. In that case, there should be mechanisms to minimize exposure to humans and ecosystems. At the same time, research should focus on identifying safer replacements.

Unfortunately, in the United States and much of the world, regulations dictating chemical use do not work this way. As pointed out in the same *Politico* article, "citizens may assume that the government thoroughly tests chemicals sold in the United States for safety, but the system largely works the other way: Chemicals are assumed safe until proved otherwise. As a result, the onus is generally on the government to prove that a chemical is dangerous before it can be regulated."[8]

Ironically, while concerns about chemicals were the trigger for the SDWA, it turns out that the number of chemicals found in drinking water sources has only grown.[9] The SDWA in practice requires drinking water facilities to meet the MCLs of a set of contaminants listed by the national primary drinking water standards. But the SDWA has not kept up with evaluating the risks of new chemicals identified in drinking water sources. Since 1996, the EPA has not added any chemicals to the regulated list, despite an increasing number of chemicals added to the Contaminant Candidate List and the Unregulated Contaminants List. Industry, agriculture, wastewater effluents, activities like oil and gas extraction, and runoff from streets result in solvents, pesticides, pharmaceuticals, and personal care products in our drinking water sources. The challenge is that human and ecosystem health impacts are unknown for many of these chemicals.

With the regulatory process too slow to protect against emerging threats, competing interests, and partisan politics in environmental regulations, is "safe drinking water" an unattainable goal? Since water is the universal solvent and bodies of water will be exposed to anthropogenic and natural activities, is it even realistic to expect drinking water to be safe? If we continue our current regulatory approaches of not acting until threats from chemicals like perchlorate and PFAS are identified, it will be challenging to keep drinking water safe. We have ample unfortunate examples of chemicals identified as hazardous to humans and ecosystem health, yet regulatory responses either do not exist or are slow and weak to respond. So perhaps what is required is a fundamental shift in regulatory decision-making, which places human and ecosystem health at the center. The precautionary principle offers the framework for such a change in regulatory decision-making.

THE PRECAUTIONARY PRINCIPLE

The precautionary principle can be considered a "better safe than sorry" approach—in other words, err on the side of caution. Suppose existing evidence is suggestive but not yet definitive that a policy or product may have a negative impact. In that case, the precautionary principle advises avoiding implementing the policy or limiting the use or manufacture of the product until there is clear evidence that demonstrates no or minimal harm.

In environmental regulations, the precautionary principle values human and ecosystem health as much as, if not more than, economic gains from a commercial product or process.[10,11,12,13] This principle states: "Where there are threats of serious or irreversible damage, lack of full scientific certainty shall not be used as a reason for postponing cost-effective measures to prevent environmental degradation."[14] In other words, if scientific data suggest that a chemical or product or practice is toxic to or can damage the environment, there should be regulatory action banning or limiting use or exposure. Decisions to use a chemical or product or implement a practice must wait until scientific data and consensus demonstrate minimal impact. The potential consequences of not heeding this principle are more severe health, environmental, social, and economic damage than the financial gains from the chemical or product or practice.

Advocates argue that the precautionary principle provides a higher level of protection to humans and ecosystems than current approaches. Such an approach would demand drinking water regulations for PFAS and perchlorate and strict regulations on their use and likely bans. It would have banned the use of lead in products far earlier than was the case. Application of the precautionary principle might have protected the millions of children exposed to lead who then struggled with developmental challenges that impacted their lives.[15] On the other hand, critics argue that the precautionary principle is too restrictive and could result in more damage than good. For example, suppose a chemical or product has demonstrable benefits but cannot be used or manufactured until data confirm it to be benign. In that case, any gains—financial, social, health—from using this chemical or product over this waiting period will be lost. In reality, the precautionary principle is not as authoritative as either advocates or critics claim, and it raises legitimate and vital questions.[16,17,18,19,20] At the heart of this debate is the process of science.

Scientific data are not definitive. There is always a degree of uncertainty associated with measurements. This uncertainty dictates the interpretation and application of data. For example, it is well established that rising levels of greenhouse gases are causing an increase in Earth's average surface temperature. Yet, science cannot precisely determine how many degrees warmer Earth will be if greenhouse gas levels double but can project a range. This uncertainty on the exact degree of warming of Earth should not override the data that indicate a warmer Earth will lead to severe global risks. But decisions on which actions to implement to reduce the warming of Earth are influenced by this uncertainty.

While scientific data must inform policies and actions, the uncertainty in data makes it challenging to make decisions that balance the risks and benefits of potential solutions. Complicating matters is that the process of science, which includes the identification of a problem, conducting research, independent verifications, and debating results, takes time. This uncertainty in scientific data, and the time needed to reach scientific consensus, is often used by industry and other special interests to prevent regulatory action. Governments wait for a scientific consensus on the evidence of harm. While waiting for this consensus, environmental and public health is at risk, and governments' efforts often address the aftermath of exposure rather than prevention in the first place.[21]

An example of the precautionary principle in practice is introducing a new drug or medical treatment. Studies measure the efficacy of a drug or treatment compared to conventional approaches. Evaluation of risks and side effects is just as critical. A shift toward this new drug or treatment is not allowed if "harm cannot be ruled out with sufficient certainty" and if "there are assumed to be greater risks involved in moving away from the status quo than in staying there. The precautionary principle puts a hurdle in place to reduce the probability of harm from treatment."[22]

Applying the precautionary principle in medical decisions seems essential as the treatment directly impacts the patient. But isn't this also the case for the water we drink? So, how can the precautionary principle be applied to lower the chance of chemical contaminants in our drinking water?

The European Union (EU) and the United States have made recent changes in regulating chemicals that attempt to embrace the precautionary principle. In 2007 the EU adopted the Registration, Evaluation, Authorisation, and Restriction of Chemicals (REACH) law. REACH applies the precautionary principle by placing the burden of proof on companies to demonstrate a chemical's safety before it is manufactured or used in a product. REACH also requires manufacturers to communicate any risks associated with the product.[23]

In 1976 the United States began regulating chemicals by passing the Toxic Substances Control Act (TSCA). This act gave the EPA authorization "to require reporting, record-keeping and testing requirements, and restrictions relating to chemical substances and/or mixtures."[24] In practice, this act has not been successful in regulating the use of chemicals. According to the Environmental Working Group, a nonprofit, nonpartisan advocacy organization, this "law was broken from the start, grandfathering thousands of chemicals already on the market" and "so weak that the EPA could not even ban asbestos, a cancer-causing substance."[25] Concerns about the law's failures to regulate and minimize exposure to toxic chemicals led to a bipartisan overhaul of this law. In 2016 the Frank Lautenberg TSCA was passed. The precautionary principle frames this act. As indicated in aspects of this act listed below, the focus is the protection of human health:

- "Mandates safety reviews for chemicals in active commerce.
- Requires a safety finding for new chemicals before they can enter the market.
- Replaces TSCA's burdensome cost-benefit safety standard—which prevented the EPA from banning asbestos—with a pure, health-based safety standard.
- Explicitly requires protection of vulnerable populations like children and pregnant women.
- Gives the EPA enhanced authority to require testing of both new and existing chemicals.
- Sets aggressive, judicially enforceable deadlines for the EPA decisions.
- Makes more information about chemicals available, by limiting companies' ability to claim information as confidential, and by giving states and health and environmental professionals access to confidential information they need to do their jobs."[26]

It is too early to see the impacts of regulations like the 2016 TSCA and REACH. These regulations are steps toward a future where, due to relying on the precautionary principle, levels of toxic chemical contaminants in water sources may be lower than today. But a public interested in a safer environment should watch how such laws live up to their objectives and ensure that current or future administrations do not weaken them. A case in point is the ongoing challenges in regulating PFAS (discussed in chapter 6). As a result of growing scientific evidence of the health risks associated with PFAS, in 2015, the EPA proposed rules to regulate this class of compounds. In 2020, the EPA revised these proposed regulations.[27] Environmental organizations and attorneys general in 18 states raised concerns that these proposed changes are weaker than the 2015 proposal. As discussed in chapter 6, the EPA has proposed action plans to address PFAS contamination more recently. In 2020, five European nations announced plans to develop a proposal that will restrict the production and use of PFAS in the EU under the REACH law.[28,29] A full proposal by these five nations to the European Chemicals Agency (ECHA) is expected to be submitted in 2022.

We drink water to keep us alive as our biological functioning requires an aqueous medium. However, the nature of the chemicals present in this water can impact our health; it can also have serious consequences for a fetus in a mother's womb. Water's chemical properties that make it essential for life and easily contaminated make applying the precautionary principle vital in ensuring safe drinking water for a community or nation. In the following two chapters, we will explore two approaches to drinking water management that draw from the precautionary principle.

In chapter 8, we will look at the role of protecting watersheds from getting contaminated in the first place, exemplifying the precautionary principle. Given water's inherent chemistry, protecting a water source from contamination minimizes the cost of treatment and protects human and ecosystem health. Watershed protection relies on natural filtration processes to maintain water quality and quantity. We will also look at the role of green infrastructure for places that may not be able to establish watershed protection programs.

Chapter 9 discusses the growing interest in potable reuse systems. In these systems, the source of drinking water is wastewater. Using wastewater might seem opposed to the precautionary principle. However, as discussed in chapter 9, acknowledging the low quality of source water dictates much more stringent treatments and monitoring than conventional water treatments. These systems are climate-resilient and can be more sustainable approaches to drinking water management. As a result, potable reuse is also a precautionary approach to the foreseeable future for many cities across the globe.

Additionally, chapter 10 focuses on the challenges faced by over 25 percent of the world's population (over 2 billion people) who do not have access to safe drinking water. This chapter looks at a range of appropriate, point-of-service solutions for rural, often low-socioeconomic communities with limited access to infrastructure. We will look at the successes and challenges of these approaches and the role of national governments in stepping up to honor the right to safe drinking water for all.

8 Protecting Nature

ECOSYSTEM SERVICES FOR DRINKING WATER

As we now know, water in the natural bodies we rely on for drinking water—rivers, lakes, streams, and groundwater—will not be pure. Humans do not need pure water, but we do need safe water. Given the chemical nature of water, the precautionary principle would argue that drinking water sources must be protected from pathogens and chemical pollutants. So, how can this be achieved?

In the early 1900s, experts debated the merits of protecting drinking water sources versus the construction of drinking water treatment systems. They concluded that the technological solution would be more effective in ensuring "the sanitary character of the water and its physical condition as to appearance, tastes, and odors."[1] Over the years, as the US Congress passed laws like the 1972 Clean Water Act (CWA) and the 1974 Safe Drinking Water Act (SDWA), it appeared that enforcement of these laws would ensure the safety of drinking water. With growing numbers of chemicals detected in water sources and outbreaks of microorganisms like *Cryptosporidium parvum*, amendments were made to strengthen the SDWA. The 1996 amendment added the Source Water Protection Assessment program.[2] As the name indicates, this program requires water utilities to develop protocols and measures that minimize contamination of drinking water sources. However, the reality is this program does not provide sufficient authority to water utilities.[3]

A case in point is the 2014 contamination of the drinking water of residents in the Charleston area in West Virginia. This contamination resulted from a spill from a chemical storage facility of the chemical methylcyclohexane methanol (MCHM). The storage facility was located on the banks of the Elk River. This river serves as the drinking water source for the area. The chemical ended up in the Elk River about one-and-a-half miles upstream of the drinking water treatment facility's intake. The local drinking water utility became aware of the contamination only after residents complained about the water's odor, and some people experienced nausea when drinking their tap water.[4,5,6]

While studies suggest no long-term human health impacts from exposure to MCHM, the question remains as to why a chemical storage facility was permitted to be close to the intake of drinking water treatment plants. It turns out that chemical storage facilities have weaker regulations and oversight than a chemical manufacturing site. As a result, such facilities are among many contamination sources difficult for water utilities to control. Other sources of contamination include nitrate and pesticide runoff from farms, pathogens from livestock farms, and chemicals used in hydraulic fracturing.[7] As we saw in chapter 6, even when regulations are imposed, such as in the GenX manufacturing plant in Wilmington, North Carolina, chemicals leak into drinking water sources. Since utilities lack the authority to prevent contamination of drinking water sources, they continue to rely on drinking water treatments to address source water contamination. Unfortunately, these treatments are increasingly unable to keep up with emerging threats to water quality.[8]

The ease with which a drinking water source gets contaminated is not surprising. Natural bodies of water are in intimate contact with the atmosphere and soil. So, are there ways to more effectively protect freshwater sources from getting contaminated in the first place? An answer to this question is establishing a *watershed protection program*, which represents the ideals of the precautionary principle. Watershed protection moves away from relying on technological solutions of drinking water treatments by recognizing the value of natural ecosystems in protecting the quality and quantity of water in freshwater sources. The concept of watershed protection has traditional roots—many indigenous communities have always respected and protected water sources.[9,10]

PROTECTING WATERSHEDS: ECOSYSTEM SERVICES

In the 20th century, scientists began researching the role of the land in influencing the quality and quantity of the water that replenishes a water source (arguably, confirming what had already been realized and practiced by generations of communities that recognized the importance of protecting water sources). The term *watershed* describes the "land area that channels rainfall and snowmelt to creeks, streams, and rivers, and eventually to outflow points such as reservoirs, bays, and the ocean"[11] (see figure 16). Precipitation as rain or snow percolates through this land before draining into the water source.

The fields of ecology, soil science, and microbiology explored the role of the landscape in influencing the quality and quantity of water in freshwater systems. As a result of these sciences, there was a growing understanding of the role of natural systems in providing "ecosystem services." A 2005 report by the Millennium Ecosystem Assessment Board defined ecosystem services as "the benefits people obtain from ecosystems. These include provisioning services such as food and water; regulating services such as regulation of floods, drought, land degradation, and disease; supporting services such as soil formation and nutrient cycling; and cultural services such as recreational, spiritual, religious, and other nonmaterial benefits."[12] How do natural ecosystems provide such services?

Picture a forest, wetland, or mountain, unimpacted by human activity (no industry or agriculture or roads)—a natural ecosystem. The ground is soil covered by litter from plant matter, animal droppings, and animal remains. This litter is broken down and dispersed by biogeochemical cycles. As the name suggests, these involve biological, geological, and chemical processes, including the carbon and nitrogen cycles. As a result, the elements of life, carbon, hydrogen, oxygen, nitrogen, and sulfur, are recycled over the years, decades, and millennia. Geochemical processes transport compounds through the air, soil, and water. Biochemical processes alter the chemical nature of compounds. For example, bacteria con-

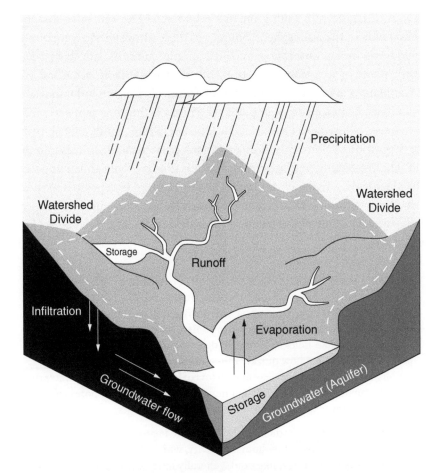

Figure 16. A watershed comprises the land through which precipitation drains into local bodies of water, including lakes, rivers, oceans, and groundwater.

Illustration by Ryann Abunuwara

vert organic compounds (carbon-containing compounds) into carbon dioxide, which is then released into the atmosphere. Photosynthesis by bacteria and plants converts this carbon dioxide into carbon-based molecules, which serve as nutrients for other life-forms. Nitrogen-fixing organisms convert atmospheric nitrogen into ammonia and nitrate, which can be taken up by plants or dissolved in water. Other organisms then break down nitrate (for example, from animal droppings) into nitrogen.

Now, imagine rain falling on this ecosystem. The rain may dissolve compounds in the atmosphere. Sources of these atmospheric compounds could be natural or anthropogenic: for example, sulfur dioxide from a coal power plant or a volcanic eruption, nitrogen oxide from a natural gas power plant or a forest fire, emissions of chemicals from an industrial site, or mercury from a coal power plant. The point is that this rainwater may have traces of dissolved chemicals. In addition to rain, winds contribute to the deposition of chemicals. Rain or wind deposit these chemicals into the soil. The ecosystem may be pristine, but it is not free from global impacts. The chemicals percolate into the soil as the material is deposited in these ecosystems either by rain or air. Soils provide physical and biochemical processing as summarized below in an article by the Soil Science Society of America:

> Capturing water is one of the most important roles that soils play in our ecosystem. This happens through the pores in the soil. The pores of a soil are important in determining if water will move into the soil and through to groundwater. Pores can be any size from small, microscopic holes in the soil to large worm-like holes and prairie dog tunnels. Soil with lots of small pores will slow down how quickly rain enters, resulting in the potential for runoff and flooding. However, soil with lots of large pores will allow water to move through quickly. An ideal soil has both large and small pores so that some water moves through, but some are stored for plants.
>
> The size of soil pores is dictated by both soil texture and soil structure. Structure can be affected by human activity, which then affects the size of the pores. Coarse textures, like sand, generally have larger pores, while fine textures, like clay, generally have smaller pores. A soil with good structure will have lots of big and small pores, even if it is clayey. A soil with poor structure, whether natural or because of erosion or compaction, will only have small pores.
>
> Land surface cover is also important. Soil covered by plants is one of the most efficient at capturing water (like grass and trees). Soil covered in concrete, like parking lots and buildings, will capture the least amount of water. Open lands that have been compacted also make it difficult for water to move into the soil.
>
> Most soils have a slight chemical charge which attracts and captures chemicals with the opposite charge. For instance, many soils are negatively charged (e.g., clayey soils). Positively charged substances, such as ammo-

nium (a form of nitrogen), are attracted to the soil. So, the soil holds the pollutants rather than releasing it into the groundwater. Negatively charged chemicals, such as nitrate (another form of nitrogen), are not attracted to the soil and may move through to groundwater.

Many pollutants are used or altered by the microorganisms living in soil. Bacteria, fungi, and more may use the pollutants and transform it to something different.

Water that doesn't enter the soil can cause floods, and it can also pick up chemicals from the land surface (such as oil in parking lots or pesticides in fields) and then enter the water supply. By increasing the amount of rain moving into the ground and by adding wetlands as buffers between land and water, we can help reduce the impact of flooding and chemicals to water bodies.[13]

As this summary indicates, the ideal soil filters water percolating through, lowers levels of suspended particles and dissolved chemicals in water, and supports a microbial ecosystem that biodegrades compounds. The nature of the plants in the ecosystem can also help purify the soil through phytoremediation, where plants sequester chemicals.[14] Plant growth and roots slow down the flow of water and allow more time for the water to percolate. The ecosystem's vegetation provides flood protection by slowing down the water flow and preventing rivers and lakes from overflowing. The water percolates through the soil, is purified through these processes, and ends up in rivers, streams, lakes, or aquifers, contributing to groundwater sources.

Natural ecosystems (natural infrastructure) can provide higher water quality and quantity than drinking water treatment plants (built infrastructure). Table 7 compares how built and natural infrastructures manage water quality and quantity. Cost-benefit analyses indicate that investing in protecting watersheds is a better long-term investment with cobenefits.[15] The water in the source will be cleaner and the supply more reliable. Water treatment costs will be lower, and the quality of the treated water will be more consistent. Watershed protection programs boost rural economics through jobs and provide recreational opportunities. Ecosystems benefit by protecting wildlife habitats.

To benefit from the ecosystem services of forests, wetlands, and mountains, we must protect the areas surrounding drinking water sources.[16,17,18] Providing this protection is the role of a watershed protection program. A

Table 7 Comparisons of the water management services provided by natural versus built infrastructure

Water security objective	Built infrastructure	Natural infrastructure
Ensure drinking water supplies at all times, even during drought	Storage such as reservoirs and tanks, water conservation, and water-use efficiency	Varied and healthy soil composition promotes infiltration and holds moisture, releasing water during periods of low rainfall, and improving water availability
Ensure water quality by protecting against chemical and pathogenic contamination	Treatment plant that includes steps such as coagulation, membrane filtration, and reverse osmosis	Plant uptake, organic matter in soils, and microbial biodegradation lower levels of chemical and pathogenic contamination
Prevent pollution from sediment or silting of waterways as storm intensity increases	Removal of deposited and suspended sediment in water treatment plant	Forests with thick root systems and native grasslands provide erosion control which is particularly beneficial during intense storm events
Flood control by reducing flow during storm events	Dams, diversion canals, levees, reservoirs, etc.	Forest layers promote water infiltration into the soil and groundwater, provide a barrier that slows water movement, and reduce runoff

SOURCE: Text in this table has been adapted from note 15 in this chapter.

watershed protection program minimizes impacts on the surrounding natural ecosystem services from human activity and maximizes ecosystem services' benefits. Protected watersheds provide high-quality drinking water to many global cities, including New York City, Boston, San Francisco, Bogotá, Cape Town, Mumbai, Madrid, Santo Domingo, and Tokyo.[19] Below is a brief history of how New York City established its watershed protection program. In the past, New York City relied on unjust practices to access pristine "upstate" water. More recently, this city has transitioned into fairer, collaborative approaches, which involve all stakeholders to ensure that the beneficiaries of the protected watershed include residents in communities local to the watershed and New York City.

CASE STUDY: NEW YORK CITY'S WATERSHED PROTECTION PROGRAM

New York City's watershed protection program is often highlighted as a case study in valuing ecosystem services. The eight million New York City residents enjoy high-quality, unfiltered water from the protected Catskill-Delaware watershed located about 125 miles away. This watershed provides about 90 percent of the city's municipal water. The water flows to the city by gravity via an aqueduct. This water is disinfected at a facility in the Bronx and distributed to residents. The city often boasts of its ability to provide a large, urban city with high-quality water at a fraction of the capital and maintenance costs compared to conventional drinking water treatment plants. The story of how the city got here is a bit more complicated.[20,21]

In the 1700s and 1800s, New York City, as with other burgeoning urban cities, struggled with water shortages and quality. Fatalities from infectious diseases were high, epidemics not uncommon, and water shortages impacted firefighting ability.[22] In the 1800s, New York City attempted to acquire water from water sources located outside the city. The power of city officials forced New York State to grant the city rights to acquire land and water rights in Westchester and gain access to the Croton River. The city constructed a dam, which created the Croton Reservoir system, and an aqueduct to supply water to the city. Rampant overuse of water and rising population led to more water shortages. The city's first response was to expand its reach into the Westchester area by constructing more reservoirs. But even as the city grew its reach in Westchester, it became clear that even more water was needed. In the early 1900s, the city acquired land in the Catskill watershed located about 125 miles away. The next phase in the expansion was to include the Delaware River watershed. Here New Jersey and Pennsylvania raised legal challenges, which resulted in some restrictions placed on the amount of water New York City could draw from the Delaware River. However, by gaining access to the Delaware River, the city established the Catskill-Delaware watershed.[23,24]

Throughout this effort, New York City used its political weight to acquire land and water rights in these areas.[25,26] The needs and concerns of local communities were overlooked and marginalized. Rightfully so, residents were furious at the city for exerting its power and authority. The land was procured often at a much lower value, people were evicted, farms were flooded, and livelihoods impacted. These unfair and unjust practices continued through much of the late 1900s. Comments from residents reveal the toll on communities for New York City to gain access to water:

> "This was put up 30 days after the date on the notice," she said, indicating a large square cloth with fringed edges. Exposure to the elements had turned it a streaked murky gray, but the black lettering still stood out sharp and clear with the message that the property to which it was attached had been condemned by the City of New York. It was dated March 8, 1962. This notice, fastened to a pole below the house, was the only legal notification the Goodrich's received that their property would be taken by the city.
>
> Mrs. Goodrich went on to recall that "four or five men would drive up . . . and while the driver sat in the car, the others got out and told the owners they had to go. They just said it's ours."[27]

In 1989 an amendment of the Safe Drinking Water Act finally triggered New York City officials to change their unfair approaches to watershed management. This amendment included a requirement called the Surface Water Treatment Rule.[28] This rule requires that surface water be treated by filtration. Exemption from this rule is possible if a city establishes a watershed protection program that includes rigorous monitoring of the quality of the water in the protected watershed. For New York City, this meant either embarking on a capital-intensive and expensive project to build drinking water treatment facilities or establishing a watershed protection plan to protect the water from contamination. Economic assessments indicated that watershed protection would be more cost-effective. However, this time, given its notorious history, the city established partnerships with stakeholders in the Catskill-Delaware community.

In 1997, through negotiations with the community, a memorandum of agreement (MOA) called the New York City Watershed Agreement was reached. The MOA required New York City to establish an ecosystem services protection program that compensates communities around this watershed to "forego industrial and commercial development that may harm water quality."[29] As a result of this MOA, the Catskill Watershed Corporation was established, which functions "as a collective choice venue for directing City funds to watershed projects and as a venue in which representatives of the City, State, watershed communities, and environmental groups can resolve conflicts."[30]

The Catskill-Delaware watershed encompasses about 1,600 square miles. The quality of the water from the Catskill-Delaware watershed is high, and New York City delivers this water to its residents unfiltered. The city is required to file every 10 years with the EPA for a "filtration avoidance determination" (FAD), which waives the city from having to filter its water.[31] Decisions on a FAD are based on assessing whether the water in this watershed meets the national primary drinking water standards. Since the MOA, the city has successfully received a FAD, the most recent being in 2017. The city tests the water for regulated and unregulated contaminants, including PFAS, hormones like estradiol, and pesticides like aldrin. The 2021 drinking water quality report indicates that these chemicals were not detected in the water from this protected watershed.[32]

Estimates indicate that the costs for constructing a filtration system for the Catskill-Delaware watershed would have been about $8–$10 billion with a daily maintenance expense of $1 million.[33] Instead, as a result of the 1997 MOA to protect this watershed, the city has spent $2.7 billion. Conversely, the city recently completed a drinking water treatment facility to treat water from the Croton Reservoir as this reservoir was not protected. The cost was $3.2 billion, and the system is capable of providing about a third of the city's water.[34] The Croton system became operational in 2015 and currently supplies about 10 percent of the city's municipal water.

The Catskill-Delaware watershed protection story is heralded as a success story of watershed protection and valuing ecosystem services—and in some ways, it is. Protecting the watershed provides a large urban city with high-quality, safe water. But this success came at a high cost to local communities, and the hardships encountered by residents in the watershed cannot be measured by dollars. The past unfair practices are lessons learned for what not to do when establishing a watershed protection program.

Observing how the MOA operates will provide experiences for building a more equitable process. If done in fair and just ways, watershed protection programs can be "win-win" by protecting natural ecosystems and natural habitats and supporting social, health, economic, and recreational opportunities for communities.

GREEN INFRASTRUCTURE

Cities like New York expanded their reach for water to outer regions when population densities were not as high. To imagine similar projects today in many places across the globe is challenging. However, there is growing recognition that urban areas will face rising demands for safe drinking water in adequate quantities. To address this 21st-century challenge requires forward-thinking approaches to water management. Some cities are turning to "nature-based solutions" or "green infrastructure" that draws from an understanding of how natural ecosystems regulate the quality and quantity of water.[35,36] While not on the same scale as watersheds, green infrastructure can be integrated into cityscapes and farmlands.

Examples of green infrastructures include green roofs, rain gardens, infiltration trenches, and bioswales (see table 8 for brief descriptions). In essence, green infrastructure mimics the natural filtration and bioremediation processes in a forest or wetland. Soils and vegetation help capture rainwater. The water is filtered through the soils, lowering the turbidity and removing dissolved chemicals. Microorganisms break down chemicals and help lower nitrate levels and other nutrients like phosphorous. The water seeps through the soil, is retained, and lowers the flood risks and storm runoff into sewers. These same principles can be applied to agricultural land, where farm runoff impacts water quality. Establishing buffer zones with vegetation and soil lowers nitrate levels in farm runoff from entering drinking water sources.[37]

Cities like Vancouver, Portland, Stockholm, New Orleans, and Berlin and the island nation of Singapore are examples of places that have embarked on green infrastructure projects. China has established the Sponge City initiative, which "will be used to tackle urban surface-water flooding and related urban water management issues, such as purification

Table 8 Examples of green infrastructure

Green infrastructure	Description
Green roofs & green walls	Coverage of existing roofs and walls with soil and vegetation.
Rain gardens	Rain gardens are designed to collect water. The vegetation in these gardens handle moisture levels from dry to flooded. The soil composition must allow for water to permeate and infiltrate, so if necessary native soils may be augmented by adding sand or compost as needed.
Infiltration trenches	These are essentially ditches to collect rainwater from surrounding areas and have soils with high infiltration rates so that the water permeates into the ground.
Bioswales	Bioswales are introduced into areas that tend to be impermeable to water, such as parking lots or roads, and as a result tend to be linear systems and greater in length than depth. These serve a purpose similar to a rain garden by slowing and filtering stormwater. Since bioswales must account for large volumes of stormwater, the soil used must have high infiltration rates and the area allocated for bioswales must be higher than that used for rain gardens. The vegetation in bioswales must also withstand a range of moisture levels, from dry to flooded.

of urban runoff, attenuation of peak runoff, and water conservation."[38] Green infrastructure projects can mitigate some of the damage done by human activity. Benefits from green infrastructure go beyond improving water quality and quantity and lowering flood risks, enhancing the quality of air, reducing the urban heat island effect, providing habitat, lowering infrastructure costs, and creating social and recreational sites within cities.

While there are multiple cobenefits of green infrastructure projects, such efforts must occur in just and equitable ways. There are concerns that unjust practices similar to New York City's impounding of land may be used for green infrastructure initiatives. For example, in a county in Florida, eminent domain was used to take over land from Black landowners to construct stormwater retention ponds.[39] Green infrastructure projects may also gentrify neighborhoods and potentially displace communities, as was experienced by historically Black neighborhoods in Atlanta.[40]

At the same time, green infrastructure projects may be most beneficial to marginalized communities who often struggle under a legacy of environmental injustice. For example, flood-prone neighborhoods of a city are often where low-socioeconomic families reside. Toward assisting these communities, the Atlanta-based organization Southface established the Green Infrastructure and Resilience Institute, which runs the Cultural-Resilience-Environment-Workforce (CREW).[41] CREW is a free job training program in green infrastructure and storm management for people living in flood-prone regions of the city. Graduates are employed in green jobs that aim to improve the environmental resilience of their community. Other organizations such as the CREATE Initiative at the University of Minnesota assist city planners in establishing inclusive policies for greening a city. CREATE has developed tool kits for policy makers to adopt and adapt.[42] Green infrastructure can make a town environmentally resilient. Still, such initiatives must be established in socially just and inclusive ways to protect the health, social, and economic well-being of all who reside in the city.

From a precautionary principle perspective, watershed protection programs are a "prevention is better than cure" approach. Given the chemistry of water, protecting watersheds that minimize contamination of drinking water sources has multiple benefits—not just for humans consuming this water but also for ecosystems and biodiversity. Green infrastructure can regulate water quality and quantity in locations where watershed protection projects are not feasible. As argued by Sandra Postel, founder of the Global Water Project, "For a couple of centuries we've been trading nature's services for engineering services," but "those engineering solutions are no longer working as well as they once did, and their economic costs are rising. Now that we better understand how nature functions and how valuable its services are, we can blend ecology and engineering, along with the social and economic sciences, to produce more optimal solutions to our growing water problems, including worsening floods and droughts."[43]

Indigenous communities have for generations respected the delicate balance of natural ecosystems that maintain healthy environments for all life.[44] Embracing traditional methods that value natural ecosystems in maintaining the quality and quantity of the water in our drinking sources is particularly relevant when environmental regulations have not sufficiently protected wetlands and streams and prevented raw sewage from being released into rivers.[45,46,47]

9   Recycled Potable Water

From a material perspective, Earth is a closed system—the amount of matter on Earth is fixed (except for a small amount that may reach Earth as cosmic debris). As a result, the amount of water on Earth is constant, but the hydrologic cycle moves this water over different locations (see figure 17). As we saw in chapter 5, from a human consumption perspective, less than 1 percent of all water on Earth is available as fresh water in groundwater, lakes, rivers, and streams. Humans rely on this fresh water for agriculture, industry, and home use. This seemingly small amount of fresh water is sufficient to support our needs. However, the distribution of this fresh water is not even across the globe, and anthropogenic activity often depletes water sources faster than can be replenished by the hydrologic cycle. Expanding populations, rapid urbanization, climate vulnerabilities, and depleting freshwater sources place pressure on drinking water sources. The world's population is projected to rise from 7.2 billion to 9.6 billion by 2050, with urban populations expanding from 54 percent to 66 percent. Urbanization in 89 countries is expected to exceed 80 percent.[1] These pressures demand resilient, sustainable, and climate-sensitive approaches to the delivery of safe drinking water. So, is it possible to recycle water and augment the hydrologic cycle instead of drawing water from depleting sources?

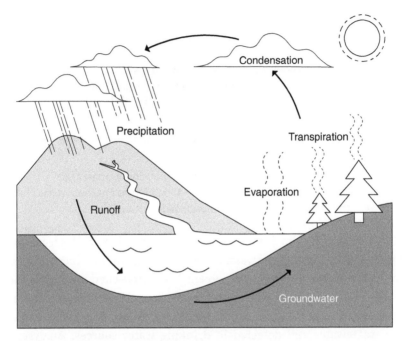

Figure 17. The hydrologic cycle, which dictates the movement of water across the Earth. This cycle replenishes freshwater sources used for drinking water.

Illustration by Ryann Abunuwara

Recycling water has an ancient history. As early as the Bronze Age (about 5,000 years ago), archaeological evidence indicates that irrigation and fertilization for agriculture used wastewater.[2] In contemporary times, agriculture and landscape irrigation, industry, and recreation use treated recycled wastewater. Given this book's focus, are there ways to use recycled water as a source of drinking water, that is, potable water? Planned potable reuse is just such an approach, the term *potable* reflecting that the recycled water is of high enough quality for drinking.

In planned potable reuse, water leaving a wastewater plant, also known as effluent, undergoes advanced water treatment. This highly treated water is delivered to end users. Why would one even consider using such low-quality, contaminated wastewater as the source of drinking water? At first glance, this might seem contradictory to the precautionary principle.

Planned potable reuse, however, can be considered an application of the precautionary principle for the following reasons:

1) Intentionally acknowledging that low-quality wastewater is the source demands that treatment methods, oversight, and assessment be more stringent than is the case in conventional drinking water treatments.
2) Investing in potable reuse systems is a climate-resilient, water-secure solution now and into the future.
3) Wastewater effluent, even if treated, is impacting human and ecosystem health. By recycling wastewater, planned potable reuse lowers the contamination of local bodies of water.

A 2017 report by the United Nations revealed that in "high-income" countries, 70 percent of the wastewater is treated.[3] In "low-income" countries, this number declines to 8 percent.[4] An often-cited number is that, globally, 80 percent of wastewater is discharged without treatment.[5] There are growing concerns about chemical contaminants being released from wastewater and detected in drinking water sources. Analyses of urban stormwater runoff reveal a "dizzying" array of chemicals, which end up in local bodies of water (a quote from the cited article states that "at this point if you can name a chemical, it can probably be found in stormwater").[6] Climate change exacerbates the frequency and intensities of stormwater runoff, adding to the chemical load delivered into bodies of water. Conventional drinking water treatments are not always effective in removing these chemicals, many of which are unregulated. Scientific evidence suggests that some emerging pollutants trigger a range of health issues, including endocrine disruption, which causes developmental and fertility disorders and cancer.[7] Some of these chemicals are also responsible for antibiotic-resistant bacteria. Tables 9 and 10 list examples of chemicals and pathogens in wastewater, respectively.[8]

Today, the stark reality is that much of the drinking water delivered by a municipal drinking water system already falls under the category of "unplanned reuse" or "de facto reuse." Unplanned or de facto reuse is when wastewater discharges into a drinking water source.[9] As seen in figure 18, water used in homes, businesses, and industries is sent to wastewater

Table 9 Examples of chemicals that may be present in wastewater and their potential sources

Type of chemical	Examples	Potential sources
Heavy metals	Cadmium, copper, chromium, lead, mercury, nickel, silver, arsenic	Industrial discharges, natural sources, water/wastewater, pipes and fittings
Inorganic chemicals	Fluoride, nitrate, nitrite, ammonia	Agricultural runoff, wastewater, human and animal waste, natural sources
Synthetic industrial chemicals	Plasticizers, biocides, epoxy resins, degreasers, polymers, polychlorinated biphenyls, phthalates	Widespread commercial use, industrial discharges
Pesticides	Household, garden, and agricultural pesticides	Domestic, agricultural, and industrial discharges
Pharmaceuticals	Nonsteroidal anti-inflammatories, antibiotics, antihypertensives, statins, veterinary pharmaceuticals	Pharmaceuticals and metabolites excreted by people and animals, domestic disposal of unused pharmaceuticals, discharges from manufacturing sites
Steroidal hormones (estrogenic and androgenic)	Estradiol, estrone, estradiols, testosterone	Human and animal waste (particularly from feedlots), excretion of natural hormones and contraceptive medication
Personal care products	Fragrances, cosmetics, antiperspirants, moisturizers, soaps, creams, whitening agents, dyes, and shampoos	Household use
Antiseptics	Triclosan, triclocarban	Household and commercial use
Per- and polyfluoroalkyl substances	Perfluorooctanoic acid, Perfluorooctanesulfonic acid	Household products (e.g., water and stain-resistant compounds, including furnishings and nonstick coatings for cookware), firefighting foams, industry
Nanomaterials	Silver, titanium oxide, zinc oxides	Used in consumer products, food storage containers, cleaning supplies, bandages, clothing, detergents
Disinfection by-products	Trihalomethanes, haloacetic acids, bromate, chlorate, chlorite, *N*-nitrosodimethylamine	Reaction between disinfectants and organic material in wastewater and drinking water (types produced dependent on source water and nature of disinfectant)

SOURCE: Text in this table has been adapted from note 1 in this chapter.

Table 10 Examples of pathogens that may be present in wastewater and the diseases they can cause

	Pathogen	Disease
Bacteria		
	Campylobacter	Gastroenteritis, Guillain–Barré syndrome
	Escherichia coli	Gastroenteritis
	Legionella spp.	Respiratory illness (pneumonia, Pontiac fever)
	Salmonella Typhi	Typhoid fever
	Shigella	Dysentery
	Vibrio cholerae	Cholera
Viruses		
	Adenoviridae	Gastroenteritis, respiratory illness, eye infections
	Astroviridae	Gastroenteritis
	Caliciviridae	Gastroenteritis
	Hepeviridae	Infectious hepatitis
Protozoa		
	Cryptosporidium	Gastroenteritis
	Giardia	Gastroenteritis
	Naegleria fowleri	Amoebic meningitis
Helminths		
	Ascaris	Abdominal pain, intestinal blockage
	Taenia	Abdominal pain

SOURCE: Text in this table has been adapted from note 1 in this chapter.

treatment facilities. The treated water discharges into local bodies of water, such as lakes and rivers, or into groundwater, which also serve as drinking water sources. For example, in the United States, rivers such as the Mississippi, Trinity (Texas), Schuylkill (Pennsylvania), and Colorado (Southwest United States and Mexico) are used as both wastewater discharge sites and drinking water sources. In the United States, for example, there are growing concerns about the chemicals not addressed by wastewater treatment. Runoff from farms, including nitrate and pesticides, is not treated and flows directly into water sources. While rivers and lakes may provide a "dilution" effect (the much larger volume of water in a river or lake dilutes the discharged water), in some cases, such

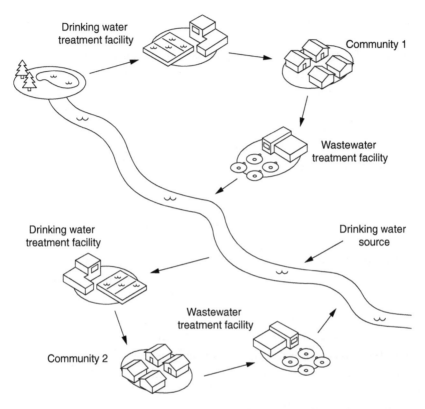

Figure 18. Schematic of "de facto" water reuse. Consumers' (homes, industry, etc.) wastewater is treated. The wastewater treatment facility discharges into a local freshwater source that serves as a drinking water source for a downstream community. A drinking water treatment facility downstream draws in this water, treats it, and delivers it to the end user.

Adapted by Ryann Abunuwara from World Health Organization, *Potable Reuse: Guidance for Producing Safe Drinking Water* (Geneva: World Health Organization, 2017).

as the Trinity River, the water often comprises primarily wastewater discharge.[10,11]

As we will see below, in planned potable reuse, the fact that wastewater is the source of drinking water for consumers translates into rigorous designs and stringent monitoring and oversight, exceeding conventional drinking water treatments.

PLANNED POTABLE REUSE

The idea of using wastewater for potable purposes is not new. One of the first attempts to do so was in Chanute, Kansas, in the late 1950s.[12,13] A severe drought in this region depleted the local water source, the Neosho River. During this drought, the water flowing in this river was essentially effluent from the sewage plant. As a result, the local government decided to treat the wastewater for potable use. The treated water was deemed safe, but this water developed a "pale, yellow color" and an unpleasant taste and odor over time. Once the drought was over, public concerns about the water's aesthetic quality triggered the shutdown of the potable reuse system.[14] Yet, this project demonstrated the possibilities and challenges of recycled water for potable reuse.

Planned potable reuse falls under two categories: indirect potable reuse and direct potable reuse. Figures 19 and 20 highlight the differences. Both rely on treated wastewater, which undergoes advanced water treatment and yields high-quality water. In indirect potable reuse (IPR), this high-quality water is transported to an environmental buffer by injecting the water into groundwater aquifers or replenishing lakes or rivers used as drinking water sources. In this way, the treated water adds to a drinking water source's capacity, augmenting the natural hydrologic cycle. In direct potable reuse (DPR), the water from the advanced water treatment facility is introduced directly into the existing drinking water system.

The significant distinction between the two potable reuse approaches is the reliance on the environmental buffer in IPR. This environmental buffer allows for natural processes to assist in the degradation of any residual contaminants that may be present in the water, dilution of these residual contaminants, and a built-in time delay to protect against any possible errors in the treatment method. The environmental buffer also helps overcome a psychological barrier for people who may be resistant to drinking recycled water from a wastewater treatment facility and uncomfortable with the perceived "toilet to tap" approach of direct potable reuse.[15]

Table 11 lists examples of indirect and direct potable reuse systems (IPR and DPR) across the world. Currently, potable reuse systems are located in drought-prone regions of the world and areas where a growing population adds pressure to existing freshwater sources. Water scarcity is

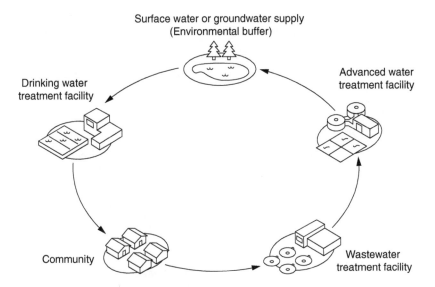

Figure 19. Schematic of indirect potable reuse (IPR). The wastewater from consumers is sent to a conventional wastewater treatment facility. The treated water is then sent to an advanced water treatment facility. The high-quality water from this advanced treatment is sent to an environmental buffer, often the local drinking water source. This water is then drawn into a conventional drinking water treatment facility for delivery to end users, repeating the cycle.

Adapted by Ryann Abunuwara from World Health Organization, *Potable Reuse: Guidance for Producing Safe Drinking Water* (Geneva: World Health Organization, 2017).

a challenge facing many countries and cities and will increasingly become so with population rise, overuse of water, inadequate water conservation efforts, water pollution, and climate change.

Using wastewater as the source for potable reuse demands sophisticated treatment methods, high technical expertise, and financial investments.[16] Places that have implemented potable reuse systems tend to be large cities with funding and technical expertise. These resources are essential for establishing complex infrastructure and consistent and proactive oversight, assessment, monitoring, and management. There is no room for system malfunction—a single error can significantly impact public health. For some of these sites, the cost of importing water, compounded

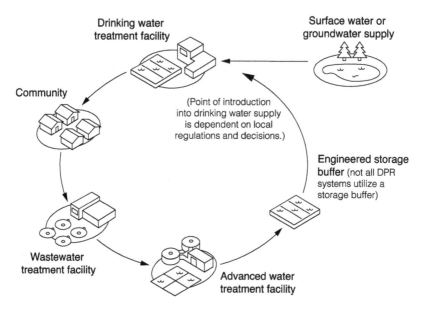

Figure 20. Schematic of direct potable reuse (DPR). The wastewater from consumers is sent to a conventional wastewater treatment facility. The treated water is then sent to an advanced water treatment facility. This advanced treatment introduces high-quality water into the drinking water system, repeating the cycle.

Adapted by Ryann Abunuwara from World Health Organization, *Potable Reuse: Guidance for Producing Safe Drinking Water* (Geneva: World Health Organization, 2017).

by growing demands, has become prohibitive and has made the transition to reuse systems economically profitable in the long run.[17] By recycling water, potable reuse is a long-term, resilient solution to climate change.

Current potable reuse sites serve as case studies for other locations to learn from and adapt to local needs as water sources become depleted or highly polluted. Advances in scientific research and analytical tools allow quantifying extremely low concentrations of dissolved compounds and microorganisms, and sophisticated, sensitive sensors allow real-time monitoring. These advances extend the work of early adopters and make potable reuse an increasingly viable and necessary strategy for a rapidly growing, urbanized world.

Table 11 Examples of indirect and direct potable reuse systems across the world

Location	Type[a]	Environmental buffer (for IPR)	Start date
Montebello Forebay, Los Angeles County, CA, US	IPR	Groundwater	1962
Old Goreangab plant, Windhoek, Namibia (replaced, see below)	DPR	—	1969 (to 2002)
New Goreangab plant, Windhoek, Namibia	DPR	—	2002
Water Factory 21, Orange County, CA, US (replaced, see below)	IPR	Groundwater	1976 (to 2004)
Groundwater Replenishment System, Orange County, CA, US	IPR	Groundwater	2008
Upper Occoquan Service Authority, Fairfax County, Virginia, US	IPR	Surface water	1978
Langford Recycling Scheme, Chelmsford, UK	IPR	Surface water	1997
Torreele, Wulpen, Belgium	IPR	Groundwater	2002
NEWater, Singapore	IPR	Surface water	2003
George, South Africa	IPR	Surface water	2009/2010
Beaufort West, South Africa	DPR	—	2010
Big Spring, Texas, US	DPR	—	2013
Beenyup groundwater replenishment scheme, Perth, Australia	IPR	Groundwater	2016
Pure Water, San Diego, CA, US	DPR	—	2035

SOURCE: Text in this table has been adapted from note 1 in this chapter.
[a]IPR stands for indirect potable reuse. DPR stands for direct potable reuse.

FROM WASTE TO POTABLE REUSE

Tables 9 and 10 make it clear that using wastewater as a source for drinking water demands extensive treatment and oversight. The concentrations of these contaminants will be higher than when drawing water from a lake or river. This realization results in stringent design, planning, construction, and oversight in water management, often lacking in conventional drinking water treatments. Unless sites can reliably and consistently provide this level of expertise and control, moving toward potable reuse of wastewater may be risky. To understand the processes involved in potable reuse, we will first look at the steps of a conventional wastewater treatment system as informed by compliance with the Clean Water Act in the United States (as mentioned above, globally, 80 percent of wastewater is discharged without treatment). Next, we will look at the advanced treatment process steps, which produce high-quality water from the water leaving the wastewater facility.

STEPS IN CONVENTIONAL WASTEWATER TREATMENT

The wastewater passes through a screen, removing large objects such as wood, plastics, bottles, and other debris. The next step is grit removal, which removes fine particles like sand. Following this pretreatment, the water undergoes a three-step process:

> Primary treatment: In this step, the water sits in containers called settling tanks. The material that settles at the bottom—called primary sludge—is pumped away. Material floating on top, like grease and oil, is skimmed off and sent to digesters to be broken down by bacterial processing.
>
> Secondary treatment: The water from the primary settling step moves to the next stage, where bacteria decompose organic matter and dissolved chemicals, including biodegradable organic compounds, nitrate, and phosphate. These bacterial processes are similar to what takes place in natural environments in soils and lakes. However, in wastewater treatment, these processes are sped up by aerating the water to allow for higher oxygen levels, which increases the rate of metabolic breakdown of organic matter and chemicals.
>
> Tertiary or advanced treatment: The water that leaves the secondary treatment step is filtered to lower levels of residual suspended particles. The water is disinfected and aerated. The aeration step ensures that the levels of oxygen in the treated water meet regulated standards to not affect the dissolved oxygen levels in the body of water into which the discharged water will flow. After this stage, the water should meet standards for wastewater discharge before being released into local bodies of water. In potable reuse, this water enters the advanced water treatment facility.

STEPS IN ADVANCED WATER TREATMENT FOR POTABLE REUSE

The water that leaves the wastewater treatment facility undergoes a series of advanced water treatment steps that produce nearly pure water. While specific steps vary, advanced water treatment may include the following steps:[18,19,20]

- Biological treatment: Any residual organic matter and nutrients like nitrate are broken down by bacteria. Examples of such treatments include biologically active media filters, anaerobic denitrifying filters (specifically for the removal of nitrate and nitrite), and soil-aquifer treatment (this step mimics the natural processes of soils discussed in chapter 8).
- Adsorption by granulated activated carbon: The term *aDsorption* is distinguished from the more familiar term *aBsorption*. Absorption happens when a sponge is wet—the entire sponge absorbs water. Adsorption is a surface-level process where dissolved organic compounds in the water are attracted to the surface of the pores in the granulated activated carbon. The dissolved organic compounds attach to the activated carbon's surface as the water runs through this filtration material. As a result, the levels of these compounds are lowered as the water filters through. Activated carbon is also used in home-based filters, visible as the black particles in the filter and why the first couple of batches of water collected may appear black.
- Low-pressure membrane filtration: In this step, membranes with microscopic pore sizes remove bacteria, protozoa, and viruses. In microfiltration (MF), the filters' pore size is on the order of 0.1 to 0.2 microns (1 micron = 10^{-6} meters). In ultrafiltration (UF), the pore sizes range from 0.01 to 0.05 microns.[21] The term *low pressure* refers to the pore sizes being small enough to require some pressure, although lower than the next stage.
- High-pressure membrane filtration: In this step, semipermeable membranes are used, which allow for the selective passage of water molecules but prevent the passage of larger dissolved compounds and ions. In nanofiltration (NF), the pore size is 0.001 microns. In reverse osmosis (RO), the pore sizes are down to 0.0001 microns. The tiny pore size requires high pressure to push the water through the membranes, so this step is energy intensive. Reverse osmosis removes most dissolved compounds, although some organic compounds like 1,4-dioxane may

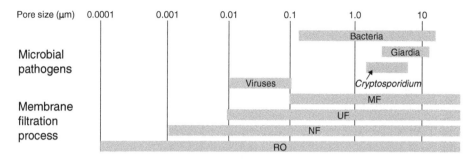

Figure 21. Pore sizes of different filtration steps to address a range of biological and chemical contaminants.

Adapted by Ryann Abunuwara from World Health Organization, *Potable Reuse: Guidance for Producing Safe Drinking Water* (Geneva: World Health Organization, 2017).

pass through these membranes (RO lowers the levels of 1,4, dioxane but does not remove it completely).[22]

- Advanced oxidation: In this step, ozone, hydrogen peroxide, and UV radiation break down organic compounds such as 1,4-dioxane not removed in an earlier step. This step also serves to disinfect by eliminating any residual pathogens.

Figure 21 shows how the pore sizes of the sequential membrane filtration steps address biological and chemical contaminants that may be present in the incoming water. Figure 22 shows the sequence of steps for the advanced treatment stages.

The sequential steps are designed to be a multibarrier approach as contaminants not addressed in a previous step are removed in the following steps. Building in this multibarrier sequence of steps in the advanced treatment is an example of the precautionary principle in action. Through an understanding of the biological and chemical nature of potential contaminants, measures are put in place to address all possible contaminants. Figure 23 is an example of how this multibarrier approach with sequential steps addresses all contaminants present in the water. Following the

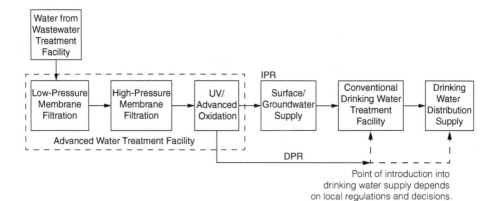

Figure 22. Schematic of the multibarrier approach. Sequential treatment steps ensure the removal of all contaminants in the recycled water. The figure shows the distinction between indirect potable reuse (IPR), where the treated water is sent to an environmental buffer (aquifer or surface water reservoir), and direct potable reuse (DPR).

Illustration by Ryann Abunuwara

path of Contaminant 3 through these series of steps, the levels decrease after the reverse osmosis step. But it is only after the final stage of advanced oxidation that levels of Contaminant 3 lower to undetectable levels. As a result, the quality of the water that leaves the advanced treatment is almost pure water and surpasses the quality of the water treated in most conventional drinking water treatment facilities.[23]

While vigilance and constant monitoring of the treated water are crucial, this multibarrier approach is designed to provide the broadest possible protection against biological and chemical contaminants, whether or not these contaminants are regulated by drinking water standards.[24] For example, studies show that activated carbon and reverse osmosis effectively lower levels of PFAS.[25,26,27] Further, the advanced oxidation steps can degrade PFAS, which means that the waste streams from the multi-step treatment systems do not contain high concentrations of these compounds.[28] Another emerging contaminant of concern is microplastics detected in natural bodies of water across the globe. Studies assessing the extent to which conventional drinking water treatment lowers levels of microplastics reveal varying results in terms of efficacy.[29,30,31,32] More

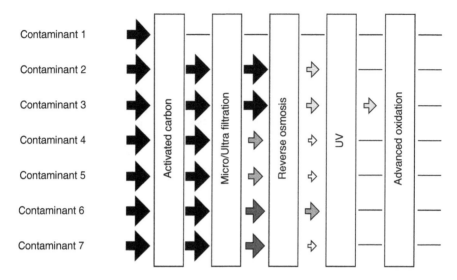

Figure 23. An example of how sequential treatment steps in the multibarrier approach address chemical and biological contaminants so that the water that leaves the facility is high quality.

Adapted by Ryann Abunuwara from World Health Organization, *Potable Reuse: Guidance for Producing Safe Drinking Water* (Geneva: World Health Organization, 2017).

studies are needed to determine the optimum conditions for conventional drinking water treatments to address microplastics in water sources. That said, the multistep advanced filtration system's ability to address contaminants ranging from the micron to the submicron scale (as indicated in figure 21) lowers microplastics in drinking water to unquantifiable levels.[33,34,35,36]

After the advanced water treatment steps, sites that utilize potable reuse as IPR send this treated water to an environmental buffer (in figure 19, this is the role of the surface water or groundwater reservoir). In a DPR system, the treated water may enter directly into the drinking water system or be mixed with water from a local source. If it is the latter, the mixed water is first treated in a conventional drinking water treatment facility before being delivered.

By explicitly acknowledging the low quality of the source water, the advanced treatments, multibarrier approach, and stringent protocols and oversight are designed to prevent failures. As a result, potable reuse can be considered an application of the precautionary principle. The water leaves the facility of higher quality than most conventional drinking water treatment systems. Planned potable reuse addresses growing public health concerns from unregulated contaminants and synergistic effects of chemical mixtures that may be present in drinking water sources.[37,38] The design of potable reuse systems is intended to produce nearly pure water, which is not the case in conventional drinking water systems.

EXAMPLES OF POTABLE REUSE

In the United States, Los Angeles County established one of the first indirect potable reuse systems in 1962, where the treated water recharges the Montebello Forebay aquifer. Recent examples include the Ground Water Replenishment System (GWRS) in Orange County, California, and NEWater in Singapore. The GWRS pumps the water from its advanced treatment facility into groundwater aquifers, preventing saltwater intrusion into these aquifers. Each day, the GWRS produces enough water to support the needs of about 850,000 residents.[39]

As early as the 1970s, Singapore embarked on a plan to ensure water security. Singapore currently relies on water from Malaysia, and this agreement will end in 2061. As advanced water treatment technologies became more robust, Singapore launched its indirect potable reuse project, NEWater.[40,41] Singapore's NEWater system currently provides up to 40 percent of the country's water needs, with expectations for this to rise to 55 percent by 2060. The quality of the water exceeds the WHO and US EPA drinking water standards.[42] NEWater is used by industry and to replenish drinking water reservoirs during dry periods.[43] One industrial end user is the semiconducting industry, which demands high-quality water and prefers NEWater to the water treated through conventional drinking water treatment. Singapore considers this a testament to the quality of the water produced by its NEWater treatment system. Through a rigorous public education initiative, which includes a visitor's center, NEWater acceptance among the public is high.[44]

While the first DPR system was constructed in Chanute, Kansas, this system was terminated when the drought ended in 1957.[45] However, this system inspired another direct potable reuse system in Windhoek, Namibia. In the 1950s, the city of Windhoek realized that local groundwater sources were being depleted at rates faster than replenishment. Projected population growth made the city recognize that it would have severe water problems within a decade, which led to the decision to supplement drinking water sources with advanced treatment of wastewater. Windhoek continues to use direct potable reuse of wastewater, although the system has been upgraded since the 1960s.[46,47,48]

Windhoek's dire situation led to the successful implementation of a DPR system, which first delivered potable water in 1969. Treated wastewater was mixed with fresh water from the local dam, which served as the drinking water source. This mixed water was sent to the drinking water treatment facility and delivered to end users. A crucial decision was that the recycled wastewater did not include industrial wastewater, limiting higher levels of chemical contaminants.

Right from the start, the engineers in Windhoek implemented a multi-barrier approach.[49] Over the years, engineers, learning from their experiences in potable water reuse management, applied this knowledge to modify this DPR system. Drawing from advances in water treatment technologies and the development of sensitive analytical and sensor tools led to designing a new DPR system, which opened in 2002. International drinking water standards such as those set by the WHO, the European Union, and the US EPA informed standards that the Windhoek treatment plant had to meet. Rigorous monitoring and assessment measurements assessed the quality of the water leaving the treatment plant. Studies evaluated if there were any public health issues in the community. So far, the results do not indicate any negative consequences.[50] The Windhoek system continues to operate today and is the longest-running direct potable reuse system. Its success has been heralded by water management experts globally, and its operation is a model for other cities to adopt. In 2015, this system launched the next research and development phase to integrate new technologies that improve sensitivity and reliability and address new threats to drinking water quality.[51]

To have successfully designed and implemented a DPR system in the 1960s is impressive, given that the water treatment technologies were not as robust and advanced as they are today. However, it must be acknowledged that during this era, Namibia was a German colony known as South West Africa and under control of the oppressive apartheid regime of the South African government. The establishment of the Windhoek DPR allowed the authoritarian White regime in this city to be sustained by the drinking water produced by this system. Namibia gained independence in 1990. In 1971 Whites were a larger fraction of Windhoek's population; today, the population is 67 percent Black, 16 percent White, and 17 percent others.

After the Chanute potable reuse system shutdown in 1957, there were no further attempts to implement direct potable reuse in the United States until 2013. A drought led to Wichita Falls, Texas, implementing a direct potable reuse system. The city was permitted to implement DPR to address this urgent need. As with the Windhoek system, DPR water was blended with lake water and delivered to homes. Residents consumed direct potable water for about a year, and the community recognized that this potable reuse helped them survive the drought.[52] Now that the area has received rains and is no longer in drought condition, the DPR is being replaced by an IPR with the treated water added to Lake Arrowhead, the local drinking water source.

San Diego has embarked on a potable reuse project called Pure Water. San Diego expects that by 2035 Pure Water will provide nearly one half of its drinking water.[53] The rapidly rising costs of importing water from Colorado and Northern California provided the impetus to construct the Pure Water facility.[54] As indicated in the informational material on the Pure Water website, "the cost [of Pure Water] is estimated to be $1,700 to $1,900 per acre-foot. This equates to less than one penny per gallon. With the current cost of imported water expected to double in the next ten years, water purification will ultimately be a more cost-effective option."[55] Economic and environmental impact analyses comparing potable reuse to desalination for San Diego revealed that the cost for desalinated water would be $2,131 to $2,367 per acre-foot. Desalination would also require 50 percent more energy and result in 46 percent higher emissions of greenhouse gases compared to the planned Pure Water system.[56]

Any major decision like switching to potable reuse requires an objective analysis of both the advantages and challenges. Below we will look at some factors that will inform a city's decision to invest in potable reuse.

ADVANTAGES OF POTABLE REUSE

i) Water quality: Given the advanced water treatment steps, the multibarrier design approach, and the necessary vigilance and real-time monitoring, drinking water quality from potable reuse systems is higher than conventional drinking water systems. Studies indicate that these systems address regulated and unregulated contaminants such as PFAS and microplastics.

ii) Replenishment of local freshwater sources: The hydrologic cycle helps replenish freshwater sources, but in many locations consumption rates of water exceed the natural replenishment rates. Recycling water through potable reuse systems reduces the water withdrawal from local freshwater sources.

iii) Climate resilience: Climate change is altering precipitation patterns globally, with many communities experiencing impacts. The future projections for many locations are grim, with longer dry seasons, hotter temperatures, and lower precipitation. Potable reuse can provide some climate resiliency.[57] Also, the lower energy demands of potable reuse systems allow for a more sustainable option than desalination.

iv) Economics: Potable reuse systems are cost-effective solutions for locations that need to pump water long distances. While coastal areas can use desalination, analyses indicate that potable reuse systems are more economical. As an example, environmental justice concerns about the high cost of desalination resulted in the withdrawal of plans for a desalination project in Monterey Bay, California.[58,59] Instead, this community is looking to extend a new potable reuse system in the region, Pure Water Monterey.[60]

v) Lower environmental and health impacts from discharges: Wastewater effluents have microbial contaminants and nutrients that trigger algal blooms, and residual chemicals impact aquatic ecosystems. The high quality of water treated from potable reuse systems means much lower environmental and health impacts and costs when the water is discharged into environmental buffers.

CHALLENGES OF POTABLE REUSE

i) High infrastructure demands: Given the low water quality of wastewater, successful potable reuse demands complex infrastructure, strict adherence to protocols, monitoring and assessment, robust multibarrier approaches, high levels of technical expertise, and consistent funding. Reliable real-time monitoring of the quality of the water that leaves the advanced treatment stages is essential. In addition, it must include redundancies to make sure that any failures are detected before the water is sent to the distribution system.[61]

ii) High economic and energy costs: While potable reuse may be an economical solution in some locations, this is often not the case compared to conventional drinking water treatments. Ongoing research efforts are investigating potable reuse technologies that are cost-effective and have lower energy demands.[62,63,64] In the long term, the higher quality of the water from potable reuse systems may be cost-beneficial when factoring in the social, health, and environmental costs associated with the discharge of wastewater effluent in locations that do not reuse wastewater. Such cost-benefit analyses will be important to assess the feasibility of wider implementation of potable reuse systems.

iii) Health concerns: There have been concerns about potential long-term health impacts when communities consume potable reuse water. Epidemiological studies conducted in Windhoek have not indicated any long-term risks.[65] Other studies suggest that health risks from de facto reuse are higher than from potable reuse.[66] That said, studies assessing the potential health risks will go far in evaluating the future viability of potable reuse systems.

iv) Public acceptance and education: One of the biggest hurdles in potable reuse is public acceptance. Potable reuse has been referred to as a "toilet to tap" system. When the public was not consulted in the decision-making process, there was resistance, which in some cases led to the cancellation of planned potable reuse projects. The "yuck factor"—a sense of disgust when people realize they are consuming "used water"—influences public opinion. Studies on public perception reveal that these come from a lack of understanding of the hydrologic cycle, that most water is de facto reuse, and that treated wastewater is often discharged into drinking water sources. Efforts involving the public in decision-making, educating them on the need for potable reuse (particularly in drought-prone regions), and informing them of the rigors of the systems that produce almost pure water have been effective in gaining public acceptance[67,68,69,70]

v) Lack of trust in city/state/federal agencies: In many cities, the public may not trust local or state officials responsible for managing drinking water systems. Cases like Flint, Michigan, and Hoosick Falls, New York, where local officials did not urgently address communities' concerns about their drinking water, do not instill public confidence. Potable reuse systems demand vigilance and oversight. People in charge of potable reuse systems must assure the public that they will protect the community's health.

vi) Regulatory needs: There is a need for federal-level regulations and policies on potable reuse. While some states in the United States have developed local regulations, decisions to permit potable reuse systems are typically made on a case-by-case basis. The WHO, US EPA, National Research Council, and organizations like the American Water Works Association, WateReuse, and Water Environment Federation have developed guidelines and frameworks to assist drinking water managers.[71,72,73] For potable reuse to become a strategy in addressing water stress and climate resiliency, the need for national-level policies and regulations is imperative.

A 2013 article reviewing the state of potable reuse systems across the world summarizes the potential for potable reuse:

> Reuse systems, particularly in potable applications, include a multi-barrier treatment framework composed of advanced unit processes, and they often incorporate resiliency (i.e., ability to adjust to upsets), redundancy (i.e., backup systems), and robustness (i.e., features that simultaneously address multiple contaminants). In comparison to conventional source waters, potable reuse is often scrutinized more carefully by the water industry, held to higher water quality standards by water regulators, and tested for a wider range of chemical and microbial contaminants. Despite an inevitably higher level of initial contamination, these systems may provide a greater level of public health protection than many common water sources treated with conventional drinking water processes.[74]

With dire projections of water-stressed regions, growing population, and climate variability, we may need to rely on potable reuse to alleviate some of these stresses. The precautionary principle would argue that researching, planning, and investing in potable reuse must happen now to ensure a community's long-term health and social and economic well-being.

Shifting toward potable reuse must include public education and involvement. Water utilities should help the public understand where their drinking water comes from, the costs associated with the delivery of safe drinking water, and the need for water conservation. The public must be made aware of the value of safe drinking water in support of public health and economic development. Successful public engagement efforts in accepting potable reuse such as NEWater in Singapore and Pure Water in San Diego are cases to learn from and replicate. Perhaps we will finally live up to the words of one of the engineers of the Windhoek direct potable reuse system, Dr. Lukas van Vuuren: "Water should be judged by its quality and not its history."[75]

10 Decentralized, Appropriate Drinking Water Treatments

Most countries in the Global North report that greater than 99 percent of their residents receive "safely managed" water through municipal water systems. The WHO/UNICEF Joint Monitoring Programme for Water Supply, Sanitation and Hygiene defines safely managed water as available on-premises when needed and free from contamination.[1] Government investments in the infrastructure and regulatory frameworks discussed in earlier chapters allow the majority of residents in the Global North access to safe drinking water. For the most part, waterborne diseases are history, and, likely, many people in these nations may not be familiar with diseases like cholera, typhoid, and dysentery.

Unfortunately, many nations in the Global South have not attained this level of success. In 2020, about 25 percent of the world's population lacked access to safely managed drinking water.[2] Figure 24 breaks down what this means for this fraction of the population who have to deal with the daily struggles associated with unsafe water. The annual death toll from waterborne diseases is over 800,000, with almost 300,000 of these deaths being children under five.[3] The World Bank estimates that the economic cost from unsafe water and poor sanitation and hygiene is about

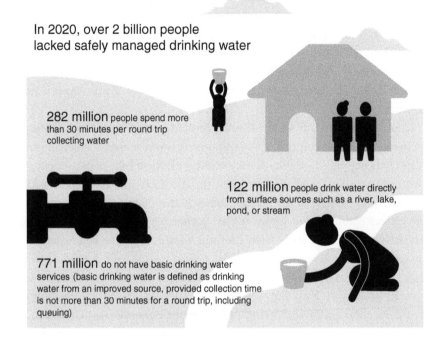

Figure 24. Depiction of the impacts on the over 2 billion people (about 25% of the global population) lacking safe drinking water.

Reproduced with permission and adapted by Ryann Abunuwara from an original work in World Health Organization, "2.1 Billion People Globally Lack Safe Water at Home" (Geneva: World Health Organization, 2015).

$260 billion annually.[4] In addition to health and economic impacts, lack of access to water takes a social and psychological toll.[5]

In 2010, the General Assembly of the United Nations (UN) formally recognized that "clean drinking water and sanitation are essential to the realisation of all human rights" (although 41 nations abstained).[6] However, ensuring this right for all residents is challenging for many countries. The economic, policy, and technical demands of drinking water infrastructure are steep. Political and economic instability and weak or nonexistent regulatory systems are additional challenges.[7,8] That being said, the number of people with access to safely managed water grew from

60 percent of the global population in 2000 to 74 percent in 2020.[9,10] These gains resulted from investments in infrastructure to provide piped water from municipal systems. However, much of these gains have been in urban areas (as opposed to rural areas) and favor people of a higher socioeconomic class.[11] Additionally, even for residents in large urban cities such as New Delhi who may receive piped water, the safety of this water is not guaranteed.[12]

Lack of access to safe water severely impacts rural communities. Many are in marginalized communities, and unsafe water makes socioeconomic gains challenging. Infectious diseases take a physical toll; costs associated with treatments, if available, add to this injustice. Children's education is impacted. Girls and women experience the brunt of the impact as they often spend many hours a day gathering water. Time spent on this daily chore takes away from the time that could have been spent on education or work, or family needs. Communities are locked into a cycle where unsafe water impairs educational, social, and economic growth, making it hard for communities to pay for water treatments. Living in rural areas and being of low-socioeconomic status makes it challenging for communities to advocate for their right to safe drinking water. Access to safe drinking water is the first step toward breaking this cycle.

DECENTRALIZED TREATMENTS

Researchers, public health practitioners, and organizations like the WHO and UNICEF advocate for decentralized or point-of-service water treatments to help rural communities access safe drinking water. These treatments are designed for small-scale implementation, a household, for example, and primarily intended for drinking. Water is fetched from a local source and treated at home. The amount of treated water produced from such systems is not by any means sufficient for the uses of tap water that we in the United States take for granted, like bathing or washing dishes and clothes. On average, a person residing in the United States uses about 76 liters (around 20 gallons) of water per shower; by comparison, a

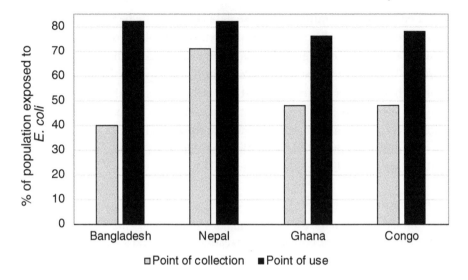

Figure 25. The percentage of people in four locations exposed to *E. coli* contamination from water collected from a source (tap, well, tanker, etc.) and after this water is stored in homes (point of use).

Reproduced with permission from WHO/UNICEF Joint Monitoring Programme for Water Supply, Sanitation and Hygiene, *Integrating Water Quality Testing into Household Surveys*, March 2021, https://www.who.int/publications/i/item/9789240014022.

biosand filter (discussed below) for a household will treat approximately 10–20 liters (around 3–5 gallons) of water in 30–45 minutes.

In most rural communities, microbial contamination in water is the primary concern.[13,14,15] The most common reason for this is poor sanitation and hygiene practices. Complicating matters is that basic hygiene practices are not possible without safe water. As a result of these interconnected challenges, the water treatment must not only address pathogens but must also be designed such that water, once treated, does not get recontaminated. For example, in some communities, a pipe or a tanker delivers water to a central location. The water, which may be safe on delivery, often gets contaminated by unclean or uncovered containers or unwashed hands dipped into a pot used to store drinking water.[16,17,18,19] The data in figure 25, collected from four different locations, reveal the challenge of recontamination. These data compare the percentage of people exposed to *E. coli* contamination from water collected from a source (tap, well, tanker, etc.) and after this water is stored in homes (point of use).[20] In all cases, there is an increase in *E. coli* at the point of use. As a result, home water treatment must not only address pathogens, but the design must also minimize recontamination. The term used for such designs is home water treatment and safe storage (HWTS).

According to the World Health Organization (WHO), a successful HWTS should address three key factors:[21]

i) Remove pathogens
ii) Be acceptable to the user—this includes aesthetic acceptance (taste, appearance, odor), ease of treatment, cost, and social acceptance
iii) Allow for widespread community adoption

The last factor is significant, as the larger the number of people consuming water free from pathogens, the lower the spread of diseases. As a result, "HWTS practices can contribute to poverty alleviation and development."[22] Below, we will look at examples of a few HWTS methods and the benefits and challenges of each. These approaches are deemed "appropriate" solutions as these are "affordable and sustainable at the household level" and rely on locally available materials and resources.[23] The success of the adoption of these treatments depends on these factors. There are decentralized, point-of-use treatments for drinking water which may yield higher quality water than the ones discussed below. However, the challenge with these "better" treatments is that they typically require materials or kits brought into the community through external organizations and have high upfront costs. Further, the technologies used in these treatment methods are unlikely to be manufactured locally and often rely on replacement parts that cannot be produced locally. The treatments discussed below can be produced locally. As a result, the improved quality of the drinking water protects the health and economy of the community, and manufacturing these treatment systems provides jobs to the community.

EXAMPLES OF DECENTRALIZED, POINT-OF-USE, APPROPRIATE TREATMENTS

i) Boiling

Since ancient times, people have used boiling to treat water. Studies have shown that heating water to a rolling boil for a minute or heating to 60°C for a few minutes is sufficient to kill most pathogens.

BENEFITS:

- Reduces levels of bacteria, viruses, and protozoa.
- Reduces incidences of diarrheal diseases.
- Easy treatment method which relies on locally available materials and existing cooking facilities.
- Relatively low-cost solution.

CHALLENGES:

- Fuel costs for boiling water can be high.
- Fuels used for cooking and boiling often contribute to indoor air pollution.
- Risk of recontamination is high (unless a narrow-mouth container stores the boiled water or a container with a tap; if the boiled water is transferred, it must be into an uncontaminated storage container).
- Treatment does not address chemical contamination. If a chemical contaminant is present, boiling may increase the concentration of this contaminant (as the water boils, it evaporates, raising the concentration [amount/volume] of any dissolved compound).
- Relies on education to ensure the treatment is effective.

ii) Chlorination

Just as large-scale water treatments use chlorination to kill microorganisms, the same approach can be applied to home treatments. Solutions of sodium hypochlorite or chlorine tablets can be purchased. Chlorine must be added in sufficient quantities to kill pathogens. An extension of this method is the combination of coagulation and chlorination. Packets that include a coagulant (like alum) and chlorine can be purchased. The coagulant allows suspended particles like soil to coagulate and precipitate; the chlorine disinfects the water. The treated water can then be filtered using a cloth.

BENEFITS:

- Reduces levels of most bacteria and viruses.
- Reduces incidences of diarrheal diseases.
- Relatively easy treatment as it does not require setup, except the purchase of the chlorinating solution.
- Relatively low-cost solution.
- The treated water may be protected from recontamination as long as there is residual chlorine.
- The coagulation plus disinfection treatment can lower levels of some chemicals (like arsenic and pesticides) that adhere to soil particles.

CHALLENGES:

- The cost of the chlorinating solution may be high for some families.
- People comment on the lingering taste of chlorine and find it unpleasant.
- If organic matter is present in the water, there is a risk of disinfection by-products (trihalomethanes) forming, which are harmful.
- Chlorination does not address contamination by some organisms like *cryptosporidium*.
- As chlorine levels decline in stored water, the risk of recontamination rises.
- Treatment does not address all chemical contaminants.
- Relies on education to ensure the treatment is effective.

iii) Solar Disinfection (SODIS)

As with boiling, relying on sunlight for treating water is an ancient practice. Water absorbs solar energy, which raises the temperature of the water. The ultraviolet (UV) radiation from the sun also kills microorganisms. The dual effect of higher temperatures and UV radiation lowers levels of pathogens. In SODIS, polyethylene (PET) bottles are used to collect water. The bottle is capped and kept in a sunny area for at least 5 hours (varies from 6 to 48 hours). In hot, sunny climates, the temperature of the water rises and can get as high as 55°C.

BENEFITS:

- Reduces levels of bacteria, viruses, and protozoa.
- Reduces incidences of diarrheal diseases.
- Inexpensive method requiring bottles available in many rural areas.
- Treatment takes place in a bottle that is capped, so the chance of recontamination is low.

CHALLENGES:

- Water must not be turbid, a term that reflects the cloudiness of water (a test for whether SODIS will be effective is that newsprint must be clear when viewed through the bottle filled with the water).
- Treatment requires hours of sunlight, which is limited during rainy seasons.
- The volume of the bottle restricts the amount of water treated, although many bottles can be treated at the same time.
- Treatment does not address chemical contamination.
- Relies on education to ensure the treatment is effective.

iv) Cloth Filter

Cloth filtration is another age-old method. Filtering water through a cloth removes suspended particles in water like soil. Since microorganisms and some chemicals adhere to soil particles, cloth filtration can help lower levels of contaminants. Cloth filtration can be effective in lowering cholera incidences.[24,25,26] Cholera is endemic in regions of Bangladesh. Surface water is a common source of drinking water and may be contaminated by *V. cholera*. Observational studies noted that cholera incidence rates were lower in families where the women used the cloth from old saris to filter the surface water into their containers.[27] High-resolution images of the sari explained why. The pores formed by the interlocking fibers of the fabric used for saris are small enough to filter out the aquatic vector (called zooplankton) which carries *V. cholera*.[28,29] The older the cloth, the more frayed the fiber is, and the network of these old fibers establishes a smaller pore size than a newer fabric. Folding the sari a few times reduces the pore size even more.

BENEFITS:

- Low-cost treatment utilizing local materials.
- Reduces turbidity and makes the water aesthetically acceptable.
- Sari cloth filtration demonstrably lowers cholera incidence rates.
- Is a common practice and socially accepted.

CHALLENGES:

- Not effective against many pathogens.
- Filters, after use, must be washed and hung out in the sun to be disinfected.
- Filtered water must be stored in containers that prevent recontamination.
- Treatment does not address chemical contamination.

v) Biosand Filter

A biosand filter is a slow sand filter system that incorporates a bioactive layer.[30] The filter container (about 1 m tall and 0.3 m wide) is concrete or plastic. Small gravel is at the bottom of the container, which prevents fine sand in a top layer from leaching out. Above the small gravel is a layer of larger gravel, which helps secure the small gravel below it. The majority of the container is filled with fine sand, and above this is a biolayer (figure 26). The biolayer is a layer of microorganisms allowed to grow on top of the sand layer. These microorganisms serve as predators of pathogens present in the water.

When water collected from a source is poured into the biosand filter, the first layer the water encounters is the biolayer. Here pathogens in the water essentially serve as "food" for the microorganisms in the biolayer. Next, as the water percolates through the fine sand, pathogens that sur-

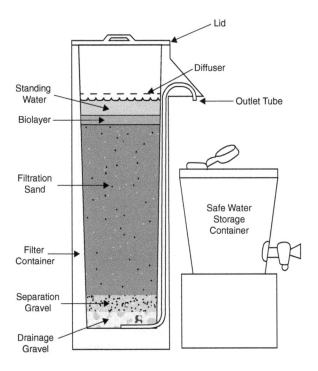

Figure 26. Construction of a biosand filter.

Adapted by Ryann Abunuwara from Centre for Affordable Water and Sanitation Technology, *Biosand Filter Construction Manual,* August 2012, https://washresources.cawst.org/en/resources/b6be2637/biosand-filter-construction-manual.

vive the biolayer get trapped in the microscopic pores of the sand. Over time these pathogens die as there is no source of nutrition. The water that filters through travels through an outlet tube and into a storage container. The storage container should be covered and have a tap to minimize recontamination.

BENEFITS:

- Reduces levels of most bacteria and protozoa.
- Reduces incidences of diarrheal diseases.
- Lowers turbidity by about 95 percent.

- Capable of producing about 12–18 liters per filtering session at a rate of 400 mL/minute (about 1.7 cups/minute).
- Uses locally available materials.
- Produced locally and manufacture of the filters provides jobs.
- Taste of the treated water is acceptable.

CHALLENGES:

- Not as effective against viruses.
- Prefiltration may be necessary if the source water is very turbid.
- Cost of a filter varies from US$10 to US$25 (while this may not seem prohibitive, people who most need these filters may have purchasing power as little as US$1.90/day).
- Filtered water must be stored in containers that prevent recontamination.
- Treatment does not address chemical contamination.
- Effective use of a biosand filter relies on education on optimal use and maintenance. The biolayer must not be disturbed. The sand filter requires regular backwashing. This maintenance may be challenging to do regularly and in water-scarce regions.

vi) Ceramic Pitcher Filter

Dr. Fernando Mazariegos of the Instituto Centro Americano de Tecnologia Industrial, Guatemala, designed a low-cost, locally sourced ceramic water filter.[31] Ceramic clay is mixed with a combustible material like rice husks or sawdust. This clay is used to throw a pot which is then fired in a kiln. As the rice husks or sawdust burns during this firing process, it creates microscopic pores of 0.2 microns in the clay. These pores are small enough to lower turbidity and bacteria.

The fired pitcher is treated with a solution of colloidal silver, known to have antibacterial properties. As the water percolates through the ceramic, the microscopic pores trap microorganisms. The colloidal silver coating helps kill microorganisms and protects the inside of the pitcher from forming a biolayer. In a ceramic pitcher filter, the formation of a biolayer must be avoided, unlike the one in the biosand filter. The biosand filter has sequential layers of filtration and subsequent layers after the biolayer address the remaining pathogens. This is not the case in the ceramic pitcher filter. Should a biolayer form along the inside of the pitcher, this can be detrimental (this is once again an example of the ease with which water can get contaminated—here by pathogens).

The ceramic pitcher can be placed in a plastic bucket to protect the ceramic from cracking. The bucket can be covered to prevent recontamination; including a tap to draw water also reduces this risk. Water passes through the filter at a rate of 1.5–2.5 liters per hour and can produce 24–48 liters of treated water a day.

BENEFITS:

- Reduces levels of bacteria and protozoa.
- Reduces levels of diarrheal diseases.
- Operates for many years.
- Easy and straightforward to use and clean.
- Lower price and easier maintenance than biosand filters.
- Uses locally available materials.
- Produced locally and manufacture of the filters provides jobs.
- Taste of the treated water is acceptable.

CHALLENGES:

- Not effective against viruses.
- Being made of ceramic, the filters can break.
- Low flow rate.
- Requires a source for colloidal silver.
- Filtered water must be stored in containers that prevent recontamination.
- Does not address chemical contamination.

vii) SONO Arsenic Filter

In some areas of the world, arsenic occurs naturally in soils and sometimes contaminates groundwater. Regions with naturally occurring arsenic in the soils include parts of the United States, Chile, Argentina,

Bangladesh, India, and China. Early in Bangladesh's history as an independent country (1971), the government was concerned about the spread of cholera from contaminated surface water. In the 1970s, the World Bank, UNICEF, and the British Geological Survey advocated that the Bangladeshi government launch a project to construct tube wells that tap into groundwater and provide cholera-free water.[32] As people in rural communities began to use this groundwater, cholera incidence rates dropped by 50 percent.[33]

Over the years, however, local health specialists began noticing a rise in arsenicosis, or arsenic poisoning. As health specialists recognized the rising levels of arsenic poisoning, Bangladeshi scientists identified the source as the groundwater from some tube wells. The international agencies who funded the drilling and installation of these groundwater wells had advised the local Bangladeshi community to drill deep into the groundwater aquifer to tap into water free from surface-level contaminations from pathogens such as *V. cholera*.[34] Tragically, this advice resulted in drilling to a depth that tapped layers of rock containing naturally occurring arsenic minerals, and as a result, arsenic ions leached into this groundwater. Despite the growing awareness of arsenic poisoning in this region, these international agencies did not raise the alarm that the groundwater might be a source of contamination even though the symptoms of arsenic poisoning were known to them and groundwater could be a potential source of arsenic.[35] Systematic and careful work by a Bangladeshi scientist identified the source of arsenic poisoning as the groundwater.[36] This work revealed the enormity of this tragedy—it was estimated that 90 million people were at risk for arsenic poisoning from the groundwater.

Abul Hussam, a faculty member in the chemistry department at George Mason University, responded to this urgent public health crisis. Hussam grew up in Bangladesh and had family living in a community where arsenic was present in the groundwater. Hussam's brother, Abul Munir, a physician in this community, was aware of the rising cases of arsenicosis in his town. The brothers collaborated on the design, development, and local implementation of the SONO filter to address this public health crisis.[37,38]

The SONO filter is a three-stage system (figure 27). In the first stage, coarse sand lies above and below a composite iron matrix made from cast iron. The iron matrix is where the critical chemistry occurs as the arsenic

DECENTRALIZED, APPROPRIATE TREATMENTS 145

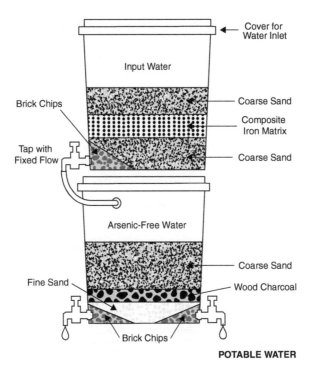

Figure 27. Schematic showing the different layers of the SONO filter. The filtered water is collected in a narrow-mouthed pot to minimize the risk of recontamination.

Reproduced with permission and adapted by Ryann Abunuwara from A. Hussam, "Contending with a Development Disaster: SONO Filters Remove Arsenic from Well Water in Bangladesh," *Innovations: Technology, Governance, Globalization* 4, no. 3 (Summer 2009): 89–102.

ions in the water get bound to the iron matrix. In doing so, the iron matrix layer lowers levels of dissolved arsenic as the water percolates. The water goes into the second stage, which includes coarse sand, wood charcoal, and fine sand. The wood charcoal helps remove some dissolved organic contaminants, and the find sand lowers the turbidity of the water. The water treated in this second stage can be drawn directly from a tap or flows into the third stage, a narrow-mouthed collection container. The filter can produce about 20 liters of water per hour. Laboratory and field tests showed that the SONO filter lowers arsenic levels to below Bangladeshi standards for arsenic (50 mcg/L) and can even meet the WHO standards (10 mcg/L).

BENEFITS:

- Lowers dissolved arsenic levels to meet Bangladeshi standards and is capable of meeting WHO standards.
- Lowers levels of other dissolved chemicals like manganese.
- Capable of producing 20–30 liters per hour of filtered water.
- Uses locally available materials.
- Produced locally and manufacture of the filters provides jobs.

CHALLENGES:

- Price of a filter is US$35–$40.
- The filter has a typical life span of five years.
- Filtered water must be stored in containers that prevent recontamination.
- Relies on education to ensure the treatment and maintenance are effective.

Table 12 compares the drinking water treatments discussed above in terms of effectiveness in lowering targeted contaminants and the typical cost of use. The WHO has conducted assessments of HWTS in lowering bacteria, viruses, and protozoa in the source water.[39,40] A treatment merits "targeted protection" if it addresses at least two of the three classes of pathogens. Chlorination, SODIS, ceramic pitcher, and biosand meet the standard of targeted protection, although effective treatment relies heavily on education and good maintenance practices.

An HWTS method not presented above, but one rated the highest by the WHO in lowering levels of pathogens, is the membrane filter, an example being the LifeStraw filter system.[41,42] The microscopic pore size of the filters is small enough to lower levels of bacteria, viruses, and protozoa. These filters rely on gravity and do not need electricity. While these filters provide the highest level of protection from pathogens, the

Table 12 Comparing the effectiveness and costs of the home water treatment and storage (HWTS) methods described in the text

Treatment	Effectiveness (% reduction)	Cost (US$)
Boiling	Bacteria >99 *Cryptosporidium* >99 *Giardia* >99 (if water is boiled at >65°C for more than 5 minutes)	$0.69–$0.88 monthly fuel cost for households
Chlorination (using sodium hypochlorite)	Bacteria >99 Viruses >99	$0.45–$3.29/1,000 liters (assuming 20 liters per household per day); costs vary depending on location
SODIS	Bacteria >99 Viruses >80 *Cryptosporidium* >80 *Giardia* >99 Helminths >90	The cost of PET bottles varies by location. They may be free or cost less than $0.50/bottle, with the number of bottles required depending on the size of the household.
Biosand filter	Bacteria >90 Viruses >80 *Cryptosporidium* >99 *Giardia* >99 Turbidity >87	$10–$25
Ceramic pitcher filter	Bacteria >99 Viruses >80 Protozoa >99 Turbidity >83–98	$15–$25
SONO arsenic filter	Arsenic >99	Around $40 per filter

SOURCE: CAWST. "HWTS Knowledge Base: Products and Technologies." https://www.hwts.info/products-technologies.

downside is that the filters are not locally produced and rely on external suppliers for the filter and replacement parts. There are upfront and maintenance costs, and unless there are financing schemes (such as microloans), relying on such filters is not feasible for many communities. Compared to the methods discussed above, the fact that a technologically sophisticated device is rated as the most effective highlights the challenges in making water safe for many rural and low-socioeconomic communities. A key reason is that the microscopic dimensions of biological contaminants and submicroscopic dimensions of chemical contaminants dictate the pore sizes of the filtering materials (as discussed in chapter 9) to effectively lower levels. Locally sourced materials are unlikely to have pore sizes as effective as the materials used in the LifeStraw filters.

While there are many approaches that households can use to reduce pathogens, how does a person decide which treatment to use? In some communities, treatments like boiling and cloth filtration are accepted practices.[43,44,45] The external agencies listed below have introduced HWTS to communities:

- Deep Springs International educates communities on chlorination practices.[46]
- Swiss Federal Institute of Aquatic Science and Technology, which developed and tested the SODIS method, has worked with communities across Africa, Latin America, and Asia to adopt this treatment.[47]
- AJPU Association, an indigenous, nongovernmental organization (NGO) in Guatemala, distributes biosand filters to rural communities.[48]
- Manob Shakti Unnoyon Kendro, an NGO established by Hussam and Munir, distributes the SONO filter.[49]
- Potters for Peace is an NGO of potters with a mission "to share this knowledge by training our partners to make filters in their own local, sustainable micro-enterprise."[50]
- The Centre for Affordable Water and Sanitation Technology (CAWST) is a Canadian-based organization with projects worldwide. They have developed educational materials and implemented workshops to help communities practice a wide range of water treatments.[51]

ASSESSING THE IMPACT OF HWTS ON COMMUNITY HEALTH

So, what is the reality on the ground? Are these treatments used in rural communities? Are communities benefiting from improved health gains? Public health researchers and social scientists have conducted field studies in rural communities to answer these questions. These studies include quantitative measurements of levels of microbial contaminants in water sampled pre- and posttreatment using HWTS and qualitative studies, including interviews, surveys, focus groups, and observations. On the one hand, communities that have embraced these treatments have lower incidence rates of symptoms such as diarrhea. But, at the same time, community-based interviews and surveys highlight challenges in adopting and maintaining a consistent practice.

For example, studies conducted in Guatemala, India, Vietnam, and South Africa have demonstrated the efficacy of boiling water in lowering levels of microorganisms from fecal contamination.[52,53,54,55] With the average daily income in some of these households being less than $1, such low-cost treatment options are essential for communities. Yet, the cost of fuel can be prohibitive.[56] People do not always realize they need to boil or treat water.[57,58,59] People do not always associate health issues like diarrhea with unsafe water. Even when people are aware of the need to treat the water, they do not regularly boil water.

Outcomes for HWTS, such as the biosand and ceramic pitcher filters, are also mixed. Field studies show that treatments do lower levels of target contaminants. For example, a study of biosand filters in a community in Cambodia showed that about 87 percent of households used the filter with durations ranging from six months to eight years.[60] The filter lowered levels of *E. coli* by 95 percent and turbidity by 82 percent. Homes that used the biosand filter had a 47 percent reduction in diarrheal diseases compared to households that did not use this filter. However, even in homes that did use the filter, stored water was recontaminated.

Complicating matters toward assessing the impact and adoption of HWTS is that each location is highly specific regarding social norms, educational level, economic status, and risk perceptions.[61] Reviews of studies suggest that while treatments may effectively address targeted contaminants, understanding their impacts through real-world experiences in communities is complicated, as noted in statements such as "the heterogeneity in results stems from the study conditions and site-specific differences."[62] While there is an agreement with the WHO's advocacy of safe storage and treatment, "it is critical to assess whether households will consistently and correctly protect their household water as there are many points for potential non-compliance" and further that "not all household-level treatment options have been found to be acceptable or desirable and few are likely to be sustainable."[63]

While the research literature reveals mixed outcomes in measuring the impact and long-term sustainability of HWTS, common factors limiting widespread adoption include the following:

i) Perception of water quality: As discussed above, communities' perceptions and understanding of water quality and its relation to health are varied.[64,65,66,67] Relying on aesthetic attributes as indicators of water quality is not reliable. People assume that water from a pipe is safe even though tests reveal otherwise. Often, people do not recognize that water quality may impact their health.

ii) Economics: Many people do not have the economic means to treat water.[68,69,70] Even boiling and chlorination are costly for some people. Local leaders hesitate when asked if they could raise funds from the community to construct a water treatment facility.

iii) Gender roles: Women are tasked with acquiring water but have to rely on their male partners for any cost incurred for treating water.[71] Some women feel they cannot ask for money even if they believe that the water consumed impacts their child's health.

iv) Time demands: People residing in rural communities have many demands on their time. Adding complicated tasks such as the maintenance of water filters makes sustained use less likely.

v) Education: People do not recognize the role of source water protection, safe sanitation, and hygiene practices, nor do people acknowledge that the water they drink may be the cause of some of their health challenges.[72,73,74,75,76] Communities need education, financial support, and training on the technical skills required for constructing and maintaining an HWTS.[77]

Ways to address these factors must be informed by local concerns, perceptions, economic means, and social and cultural practices. Engaging communities in every stage of implementation is essential for widespread and long-lasting adoption. The Centre for Affordable Water and Sanitation Technology (CAWST) provides examples for how to achieve this, as advocated in a statement from their website:

> The most proactive way to find a sustainable solution to a community's water issues is to spend a LOT of time listening. Look at the following: people's needs and preferences; economic factors; availability of local skills and labour; environmental and technical factors; local or national regulations. Finally, look at what hasn't worked and why. Sometimes failures are the biggest lessons.[78]

CAWST's website has an extensive repository of information, along with manuals and case studies of implementations of a range of HWTS approaches. The organization's mission is to empower local communities with the knowledge and practices for HWTS and support rural development schemes through jobs in the construction and maintenance of water treatments.

Addressing the global challenge to provide access to safe drinking water for 25 percent of the population may also require new approaches. The past has been about bringing well-intentioned solutions developed elsewhere to rural communities. Are there solutions that are more in line with local practice and knowledge? A case in point is a concept paper for a water treatment strategy for a rural community in Tanzania.[79] This paper describes a conceptual design for rainwater harvesting, a local practice in Tanzania. Rainwater collected before it percolates into the ground may be less contaminated than local water sources. The collected rainwater will be treated in a facility using affordable treatments such as biochar and iron-based filters combined with slow sand filtration. Biochar is a form of charcoal made from burning biomass and is an accessible and sustainable filtering material in rural communities. Studies have shown that a combination of slow sand, biochar, and iron filtration lowers pathogens and dissolved chemicals.[80,81] This treated water is distributed to homes via pipes. Execution of a project like this will require government support and funding, as stated by the authors: "Translating the concept into reality requires a supportive policy and institutional framework, and a shift from the misconception that technologies to solve problems in Africa are to be imported from the West."[82]

Ultimately governments must step in, work collaboratively with local communities, and assess how best to provide economic, educational, social, and technical support for appropriate treatments that deliver safe, reliable drinking water. The concept paper describing the rainwater filtering system exemplifies how nations can achieve this. Successful implementations of such projects will go far in demonstrating how solutions relying on local knowledge, practice, and materials, and with support from national governments, may effectively address the global challenge in access to safe drinking water. These investments also relieve burdens on households having to treat water, something that most people in the Global North have the luxury of not worrying about.

As an example, in 2019, India launched an ambitious project "to provide safe and adequate drinking water through individual household tap connections by 2024 to all households in rural India."[83] Called the Jal Jeevan Mission (loosely translates to "water is life"), this project includes water conservation, rainwater harvesting, and water reuse strategies. The project aims to train residents in these communities to conduct water quality tests and educate them on ways to recycle and reuse wastewater. The water quality testing is a crucial component, given the data in figure 25, highlighting the challenges of recontamination after collecting water from the source. Progress toward meeting the project goals by 2024 is monitored through a dashboard that includes data on the number of households connected, water supply in schools, and water quality measurements.[84] As a nation, India bears significant health, social, and economic impacts from waterborne diseases. Every year about 40 million Indians experience illnesses from waterborne diseases. Annually this results in a $600 million loss due to medical and loss of labor costs.[85] By providing access to piped water to its rural communities, India is investing in a healthier future for its people.

Since 2019, villages have gained access to piped water. However, this is an extraordinarily complex project and not surprisingly faces challenges.[86] For example, some question the long-term sustainability of water access in water-stressed areas that will only be exacerbated by climate change. In addition, unreliable electricity generation in some rural areas means that electric pumps cannot run, and hence the distribution of water is unpredictable. And to put into place the plans to train residents in villages to maintain distribution systems, implement water reuse and recycling, and conduct water quality tests will not be easy to achieve, particularly at the scale needed in a country like India. And finally, for such a large-scale project to be successful in the long term and yield the intended benefits, it must be free from politicization, focus on equitable distribution and access, and implement sustainable infrastructure. While these challenges must be recognized, the benefits are significant if India expands access to piped water into rural communities. Women will not have to spend hours gathering water. Children will be healthier and less affected by waterborne diseases, and girls will have time to go to school.

The signing of the UN resolution declaring that safe water and sanitation are human rights must translate into meaningful action. Investing in systems that deliver safe water will reap benefits for the community and the nation. Studies show that for every $1 invested in safe water and sanitation, the return on this investment is $4.[87] Access to safe water is a necessary step toward improving hygiene practices. Safe water leads to improved health, education, and social and economic outcomes, saving millions of lives.

11 Valuing Safe Drinking Water

Ever since humans transitioned from nomadic to settled life, we have needed reliable sources of water. Unfortunately, we have not paid attention to the impacts of agriculture, industry, and domestic waste on water quality. The first answer to address poor local water quality was transporting water from pristine areas to population centers. As science advanced and identified the causative relationship between microbial contamination and the spread of infectious diseases, the response was to treat unsafe water to make it safe. Research identifying chemical contaminants as rising threats triggered government regulatory action. In the United States, laws such as the Clean Water Act and Safe Drinking Water Act were passed. As a result, residents now have seemingly unlimited amounts of safe drinking water in countries that invested in the necessary scientific, engineering, and regulatory infrastructure. Investments made have been realized manyfold by the health, societal, and economic benefits possible when safe drinking water is widely accessible.

While past successes in the delivery of safe drinking water are worth championing, the focus on treating water postcontamination rather than preventing this in the first place is now coming to haunt us. We are now in a situation where we take our access to safe drinking water for granted.

However, doing so is dangerous as absent constant oversight and vigilance, we can quickly lose the luxury of safe drinking water. Water's chemical nature is why water is easily contaminated and requires complicated and costly treatments to protect, maintain, and improve its quality. While the Safe Drinking Water Act has undoubtedly improved drinking water quality, as discussed in chapter 5, this act is by no means perfect. The cases of contamination of drinking water by lead, nitrate, and PFAs addressed in chapter 6 exemplify how challenging it is to ensure the safety of drinking water even when there are established regulatory frameworks. Serious attention and investments are urgently needed to address inequities in this access, now further exacerbated by the ripple effects of COVID-19.

As nations reimagine systems and practices in response to the pandemic's multiple impacts, a precautionary principle framework for environmental regulations should be embraced. Doing so will go far in preventing water sources from getting contaminated, allowing for positive public health outcomes, and protecting the health of ecosystems. While politicians have often portrayed environmental regulations as threats to the economy as these impose compliance costs on industry, in reality, this is not the case. Cost-benefit studies show that investments made in safe water are more than recovered due to improved health outcomes and social, educational, and economic gains. The beneficiaries of safe water are all of us, regardless of our political affiliation. Therefore, it is incumbent on us to recognize and respect the value of safe drinking water as a life-sustaining resource.

WHAT DOES IT MEAN TO VALUE SAFE DRINKING WATER?

Adam Smith's 1776 book *An Inquiry into the Nature and Causes of the Wealth of Nations* discusses the concept of the "paradox of value" by comparing the "value of use" to the "value of exchange." Smith used what is now known as the diamond-water paradox:

> The things which have the greatest value in use have frequently little or no value in exchange; on the contrary, those which have the greatest value in exchange have frequently little or no value in use. Nothing is more useful than water: but it will purchase scarcely anything; scarcely anything can be

had in exchange for it. A diamond, on the contrary, has scarcely any use-value; but a very great quantity of other goods may frequently be had in exchange for it.[1]

Smith went on to explain that the value of exchange was a measure of labor: "The real price of everything, what everything really costs to the man who wants to acquire it, is the toil and trouble of acquiring it." Through this economic lens, safe drinking water is also a value of exchange and not just a value of use. Ensuring the safety of water requires significant labor and cost. Simultaneously, since access to safe drinking water is essential for a community's health and economic well-being, it must be priced so that all can afford and benefit from it. Achieving this balance between recovering costs and affordability is tricky. When the price of safe water is low, its value is not recognized and it is often overused. Overuse of water results in higher contamination levels, further raising the cost of ensuring its safety. Are there ways to address this conundrum?

Municipal water utilities are expected to recoup the cost of treatment and delivery of drinking water from ratepayers. In 2018, averaging the cost of municipal water across the United States, the price per gallon for tap water was about $0.005.[2] As a comparison, the cost of a gallon of bottled water ranges from $0.90 to $8.26 per gallon. This price range depends on the brand and the bottled water's volume; the most common size is 16.9 ounces.[3] On average, then, a gallon of tap water is 167–1,500 times cheaper than bottled water. In many cities, water rates include the cost of delivering the water and wastewater treatments. Even after factoring in the wastewater treatment rate, municipal water is 70–600 times cheaper than bottled water—and water from the tap is more tightly regulated than bottled water.[4]

While tap water may be cheaper than bottled water, the reality is that water rates are now rising in many cities, and many people are struggling to pay their water bills.[5,6,7,8] Maintaining existing infrastructure, compliance with regulations, and upgrades to address new threats raise costs. Utilities pass these costs on to consumers. For example, cities in states like Arizona, California, and Texas, which are vulnerable to drought, have embarked on large-scale water projects. San Francisco, California, is nearing the completion of a $4.8 billion water infrastructure project to protect against earthquakes. San Diego, California, and El Paso, Texas, are investing in potable reuse systems. Phoenix, Arizona, has a $300 million water infrastructure project to protect against droughts. While these upgrades are essential to ensure water security and safety for residents, affordability is clearly of concern. Water utilities are increasingly finding themselves having to balance competing issues: "Earn enough revenue to repair broken pipes, keep water affordable for the poor, and do so while selling less of their product."[9]

Rising costs in upgrading infrastructure and meeting compliance with drinking water regulations also place an unfair burden on some communities. In large urban areas, utilities can often obtain loans or use bonds to finance upgrades to water and wastewater infrastructure. The large population in these cities allows them to recoup these costs.[10] However, there are also examples of cities unable to recover these expenses through water rates paid by the consumers that end up in debt. Some estimates suggest that the debt from paying for water infrastructure in the United States is about $1.7 trillion.[11]

Smaller, rural communities or cities where populations are in decline cannot secure financing for such infrastructure projects. Even if they did, they are unlikely to recover costs through their customer base.[12,13] Lacking finances to upgrade affects the ability of utilities to meet regulations, and as a result, these communities are experiencing higher levels of violations of drinking water standards. While the intent of federal-level regulations may be to deliver the same drinking water quality for all who reside in a nation, regardless of zip code, this has never been realized in the United States. Now, rising costs to deliver safe drinking water only exacerbate these inequities. Utilities have shut off water supplies when families cannot pay their water bills. For others, the inability to pay water bills has resulted in liens on properties and even eviction from homes.[14,15]

The reality is that delivering safe drinking water is expensive and complicated. There is growing recognition that what consumers pay through water rates does not cover the true cost of delivering this water. Many have argued that water rates for municipal water have been too low in the United States, underpricing its delivery costs and undervaluing what safe water means to a society. It has often been stated that municipal water cost is higher in nations of similar economic standards as the United

States. But making such comparisons is challenging given the social, economic, and cultural differences among nations. Below, we will look at a few examples of how some cities and states in the United States are balancing affordability with the realities of the cost of delivering safe drinking water.

Some cities rely on water pricing structures that balance the utility's financial needs and water affordability and encourage water conservation. For example, Santa Fe, New Mexico, is a city that often experiences drought. This reality triggered the city to adopt a tiered pricing model that balances affordability and water conservation. For the first 7,000 gallons of water used per month, the rate is $6.06/1,000 gallons.[16] For monthly use higher than 7,000 gallons, the rate increases to $21.72/1,000 gallons. This tiered pricing model has lowered water consumption in Santa Fe by 20 percent. At the same time, the city's population grew by 10 percent.[17] As a result of the tiered pricing model, the average water use in Santa Fe is about 160 gallons/day/household; in the United States, the average consumption of water is 400 gallons/day/household. A quote by Robert Glennon, an author of a book on US water policy, captures how this approach values safe water: "The beauty of tiered pricing is that it doesn't prevent people from using water, and it doesn't rely on government regulations. But it insists you pay more for extra water for your lawn than for basic human needs."[18] At the same time, implementing a tiered pricing system must pay attention to the socioeconomics of community members. For example, a larger family will tend to use more water and should not be penalized because of the tiered pricing structure.

Cities like Philadelphia, Baltimore, and Cleveland have introduced water pricing based on income levels.[19] These efforts respond to increasing water rates to fund upgrades to infrastructure, which can make municipal water unaffordable for some residents. Philadelphia's Tiered Assistance Program (TAP) uses a water rate based on household income.[20] A press release from the Philadelphia mayor's office states "We are committed to making basic human services affordable for all citizens of Philadelphia, and this program is designed to do just that, which benefits the city as a whole." In Baltimore, water rates are projected to increase by 9 percent to pay for upgrades to infrastructure. Local organizations such as the Baltimore Right to Water Coalition advocated that city officials pass legislation protecting residents' access to municipal water.[21] The Baltimore City Council passed the Water Accountability and Equity Act, including an income-based water billing program to protect low-income families from price hikes.[22,23] Buffalo has a program where commercial and industrial sectors subsidize the price of water for residential users. This city also has an income-based water pricing program.[24]

In 2019, the California Legislature approved a bill that established a Safe and Affordable Drinking Water Fund.[25,26,27] The primary focus of this fund is to provide safe water to disadvantaged communities whose local water systems are contaminated or who do not have access to piped water. These funds help consolidate smaller utilities, connect them to a larger neighboring system, or provide piped water to communities that do not have connections yet. This fund aims to help over one million residents in the state finally get access to safe, affordable, piped water.

These efforts at the local and state levels are commendable and are steps toward helping all residents access safe drinking water regardless of socioeconomics. However, access to affordable, safe water must not depend on which city or state you happen to live in. There needs to be more aggressive ownership at the federal level to address the many historic and new threats to equitable and affordable access to safe drinking water. Communities in places such as Flint, Newark, Hoosick Falls, and the Central Valley have tragically borne the cost of delayed and insufficient investments in infrastructure. The CWA, SDWA, and TSCA must include more stringent regulations, such as stronger protection of drinking water sources and regulating chemicals such as PFAS and perchlorate that pose risks to ecosystem and human health. So, what will push governments to take serious steps in ensuring equitable and affordable access to safe drinking water for all? In the 1950s and '60s, the public insisted on governmental response to environmental degradation. It is now time for us to demand that our political representatives act in ways that protect the right to safe drinking water for all.

PUBLIC'S ROLE IN VALUING SAFE WATER

Imagine a day without water. What might that look like for you? The Value of Water Campaign asks the public to do just this in an annual event

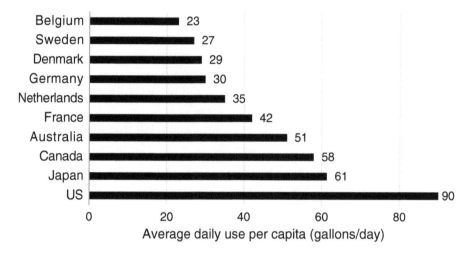

Figure 28 Data comparing average per capita household water use across different nations (2017 data for Canada and the United States; 2018 data for the other nations).

Sources: US Geological Survey, "Water Q&A: How Much Water Do I Use at Home Each Day?", June 20, 2019, https://www.usgs.gov/special-topic/water-science-school/science/water-qa-how-much-water-do-i-use-home-each-day?qt-science_center_objects=0#qt-science_center_objects.

Statistics Canada, "Survey of Drinking Water Plants, 2017," released June 11, 2019, https://www150.statcan.gc.ca/n1/daily-quotidien/190611/dq190611b-eng.htm.

International Water Association, "Total Water Delivered for Households," 2018, https://waterstatistics.iwa-network.org/graph/VG90YWwgV2F0ZXIgRGVsaXZlcmVkIGZvciBIb3VzZUhvbGRzIChtsy9jYXBpdGEveWVhcik/undefined/2018.

reminding people that such a day would mean "No water to drink, or even to make coffee with. No water to shower, flush the toilet, or do laundry. Hospitals would close without water. Firefighters couldn't put out fires, and farmers couldn't water their crops."[28] So, think for a moment about how you used water today and what you would do if your water supply suddenly stopped for a day or two. Then, with this in mind, below are some suggestions for how you can value safe water.

WATER CONSERVATION

Comparing water consumption across some Global North nations, the United States has a high per capita water consumption, as evident in figure

28. These data suggest that we could consume less water in the United States and maintain a "standard of living" equivalent to nations considered similar by economic measures. In many of these nations, the monthly cost of water for the consumer is higher than in the United States. As mentioned above, making such comparisons can be tricky, but it is worth pointing out that the price we pay for water influences how much water we use. The cheaper this water is, the less likely we are to pay attention to how much we use, undervaluing what access to safe water means. In turn, utilities cannot recover the true cost of what it takes to deliver this safe water.

Ways to lower water use include installing low-flow showerheads and toilets, fixing leaks, turning the tap off while brushing your teeth, using energy and water-efficient appliances, and reducing water used while gardening by avoiding sprinklers.[29] These may seem insignificant steps, but they add up, particularly if each person in a town or city tries to lower their water use. Using less water lowers the stress on drinking water sources and the volume of wastewater needing treatment, reducing the cost of wastewater treatment.

Reducing individual water use has other benefits as well. Much of the water infrastructure is decades old and urgently needs repairs and upgrading. High volumes of wastewater overwhelm sewer systems and wastewater treatment plants. Increasing population growth adds further stress to these aging systems. Climate change is driving more severe weather patterns, such as intense downpours. As a result, wastewater treatment systems often exceed capacity, releasing raw, untreated sewage into local bodies of water. These overflow events compromise the water quality of natural bodies of water and degrade aquatic ecosystems. In addition, such overflow events are a public health concern, particularly if this same water is used by a "downstream" community and raises drinking water treatment costs. Sewer overflow events are a challenge in cities that rely on combined-sewer systems that combine water from road runoff and home and business wastewater. During heavy downpours, these combined sewer systems often exceed the capacity of wastewater treatment plants which then release untreated sewage water into local bodies of water. Often, these utilities request residents to limit water use during downpours to lower the flow into wastewater treatment plants and avoid having to release untreated sewage water (see if this is the case in your community by looking at the website of your local drinking water system or city's environmental agency).[30,31] Be mindful of how much water you use (for example, by looking at your water bill) and consider ways to lower water consumption. Reducing your water use not only saves you money but can alleviate pressures on water infrastructure, lowering maintenance costs, and lessen environmental impacts on local bodies of water.

EDUCATION

The more we understand the process from source to tap, the more informed we will be about our actions. We have become increasingly aware of where our food comes from, but we tend to be less educated about the process from drinking water sources to our taps. Spend time on your local water utility's website to understand the process from source to tap and learn about your utility's work in ensuring the quality of the water you drink. Another way to educate yourself is to review your local water utility's Consumer Confidence Report (also referred to as drinking water quality reports). The SDWA requires every municipality to publish annual reports to inform communities about their drinking water quality.[32]

As you review these sites or reports, pay attention to your local drinking water source and if your utility protects the source from contamination. Look at the processes involved in treating your water. What are the filtration and disinfection methods used? Does your tap water meet the national primary drinking water standards? Are all contaminants in compliance? Does your town have lead service lines, and does your report discuss treatments that address lead, such as adding orthophosphate? Is there a free service to test your tap water for lead? Is your utility screening for unregulated contaminants as required by the EPA? Does your utility have plans to upgrade infrastructure to address local threats such as chemical and microbiological contaminants, adapting to climate change, or investing in green infrastructure projects? Educating yourself on your drinking water quality will help you appreciate what it takes to deliver your water.

Educate your neighbors and encourage them to familiarize themselves with your town's drinking water quality report. Perhaps there are ways you and your community can communicate concerns, questions, and suggestions to the local city council or other governing bodies that manage your town's municipal water. Some town councils have public hearings about water-related issues, so consider attending meetings. Prior to the COVID-19 pandemic, many drinking water and wastewater utilities hosted public tours of their facilities. These were great opportunities to not only learn about your local water systems but to appreciate the work of the people that protect our public health. Hopefully, such tours will resume once it is safe to do so.

UNDERSTANDING WATER RATES

Water rates have been increasing in many cities. Not surprisingly, residents are upset as their water bill increases. That said, it may be worth pausing and understanding why your rates are increasing. Is it because the water utilities must upgrade infrastructure to ensure the safety of the water? At the same time, not every rise in water rates may be justified. There have been unjust examples of towns facing financial challenges that increase water rates to generate revenue.[33] Knowing that people need water, officials have used the funds accrued from higher water rates to pay for unrelated costs.[34] All to say, it pays to explore why your city is raising water rates and assess if the reasons appear valid. Pay attention to whether there are plans to ensure that all residents can afford the water. The approaches of Santa Fe, Philadelphia, or Baltimore may be ones your city should consider adopting.

PROMOTE GREEN INFRASTRUCTURE IN YOUR COMMUNITY

A growing number of cities offer tax incentives for residential and non-residential properties to construct green infrastructure projects such as green roofs and rain gardens.[35,36] As we saw in chapter 8, green infrastructure improves water quality and controls runoff into local bodies of water. Replacing concrete with green infrastructure has other benefits, including lowering urban heat island effects, improving soil and air quality, increasing local biodiversity, and providing recreational and social spaces for communities.

CONSIDER WATER-RELATED CAREERS

If you are a student or looking for a career change, explore ways to connect your academic and professional goals to the water sector. Given that we will always need safe drinking water, jobs in this sector will be stable and will only grow. The expertise required is extensive—from science and engineering to public health policy to economics and activism—and there are opportunities for all disciplines to engage with water-related issues. Working in this field will make you integral to implementing solutions necessary to address the global challenges in ensuring equitable access to safe drinking water.

VOTE

Finally, in democratic societies, it is incumbent on each of us to educate ourselves about issues that matter to us and cast informed votes. If access to safe drinking water matters to you, use your voice to vote for political representatives who you believe will act in ways to protect this right for all who reside in your nation. It is the responsibility of those privileged enough to have safe drinking water to represent the concerns of those who do not and may not have the resources or political voice. A nation benefits maximally when 100 percent of its residents have access to safe drinking water.

IN CONCLUSION

This book explored the molecular-scale properties of water that make it both essential for life and easily contaminated. By recognizing this chemical paradox of water, it becomes evident why we must fundamentally change how we protect, manage, and use water and not take our access to safe water for granted. The precautionary principle must frame our policies and regulations regarding drinking water. We know that water in the natural environment will dissolve chemicals it comes into contact with, and microorganisms thrive in water. While not all chemicals and microorganisms are harmful, a community exposed to a chemical such as lead or PFAS or a pathogen such as *Cryptosporidium* will experience negative impacts.

While a nation like the United States has established regulatory frameworks that govern drinking water quality, we see far too many cases when communities are exposed to contaminants in their drinking water. Outdated infrastructure, declining federal funds to drinking water systems, far too slow regulatory processes in keeping up with threats, and political and industry pressures are underlying reasons for many of these contaminations. Long-standing racial and socioeconomic inequities have prevented some communities from ever enjoying the benefits of access to safe drinking water. Rising costs of drinking water management and compliance will only exacerbate these inequities and, if anything, add to the numbers of people who either lack or may not be able to afford safe water. By understanding these challenges, those of us with the privilege of this access must demand action by our political representatives to ensure that all who reside in the United States have equitable, affordable access to safe drinking water. Yes, water is life—but it is safe drinking water that not only protects public health but is essential for poverty reduction, food security, supporting peace and human rights, enabling education and gender equity, and ecosystem health.[37] To realize these rights and protections for all life on Earth, the paradoxical behavior of H_2O must shape and inform individual behavior as well as policies, regulations, and actions for how we protect, use, manage, distribute, and treat water.

In concluding this book, I end with a few quotes to reflect on as you explore water-related issues beyond those raised in this book. Such quotes remind us why we must not take our access to safe drinking water for granted.

Thomas Fuller (18th-century historian):

"We never know the worth of water till the well is dry."

Albert Szent-Györgyi (biochemist):

"Water is the most extraordinary substance! Practically all its properties are anomalous, which enabled life to use it as building material for its machinery."

Sylvia Earle (marine biologist):

"There's plenty of water in the universe without life, but nowhere is there life without water."

Autumn Peltier (member of the Wikwemikong First Nation; in 2018, at the age of 13, she addressed the UN General Assembly; these quotes are from her speech):[38]

"Our water deserves to be treated as human with human rights. We need to acknowledge our waters with personhood so we can protect our waters."

"No child should grow up not knowing what clean water is or never know what running water is."

"We all have a right to this water as we need it—not just rich people, all people."

Acknowledgments

It is unlikely I would have written this book were it not for the fact that I teach at an institution where colleagues and students engage in interdisciplinary dialogue. Being at Eugene Lang College has allowed me to think about chemistry in ways I may otherwise not have. I thank my colleagues and students for the discussions that helped my thinking in integrating across disciplines.

Thank you, Ryann Abunuwara, for your artwork and your grace, patience, and willingness to make the many corrections I asked of you. Your creative input in identifying ways to communicate information and ideas through your illustrations was invaluable.

I would like to thank the peer reviewers. While all the reviewers were positive about the book, each provided valuable comments and perspectives. I believe their feedback helped strengthen the arguments and information presented in this book, particularly since each approached the topic of drinking water from a different perspective. Given that my intent for this book is to connect across disciplines, reading their comments was extremely helpful. I hope I have successfully integrated their feedback into the book.

Thank you to the University of California Press for accepting this book for publication. I am grateful for the feedback and support from Stacy Eisenstark, Environmental Studies Editor, particularly in the early stages of writing this book. Thank you, Naja Pulliam Collins, for working with me and patiently answering my many questions via emails and Zoom calls. I would like to thank Francisco Reinking, Teresa Iafolla, Chloe Wong, the book cover design team, and other people from UC Press who were involved in the production of this book.

Thank you to Linda Gorman for her careful copyediting. And thank you to PJ Heim for creating the index.

Thank you, Sanda Balaban, Davida Smyth, and Michael Wentzel, for writing the publicity statements.

I would like to thank Lisa Maria for her insightful comments on the book proposal.

I am fortunate to have an incredibly supportive family willing to read book proposals and chapters. I would like to thank Arun, Chitra, Anu, Markus, and Suhasini, all of whom provided important and critical feedback and ideas (and support in so many ways). In addition, their professional backgrounds and personal interests were a real asset to me as they could speak to the substance of this book.

Thank you, Peri, Harold, Roy, and Marcie, for your support.

I express my deep gratitude and love to my parents. My father, I believe, would have been very happy to know I have written this book. I know he would have had many important points to discuss with me. My mother, a librarian by profession, instilled in me, at an early age, the power of books.

Thank you, Deven and Anita. I do not have sufficient words to express my deep, deep appreciation, gratitude, and love. You have both provided crucial feedback (belonging to the "target audience") and so much support in immeasurable ways.

Finally, this book is for Jay—you will always be with me.

Notes

PREFACE

1. United Nations Development Programme. "Sustainable Development Goal 6: Ensure Access to Water and Sanitation for All." Accessed July 7, 2020. https://www.un.org/sustainabledevelopment/water-and-sanitation/.

2. Michigan Civil Rights Commission. *The Flint Water Crisis: Systemic Racism through the Lens of Flint.* Report of the Michigan Civil Rights Commission, February 17, 2017. https://www.michigan.gov/documents/mdcr/VFlintCrisis-Rep-F-Edited3-13-17_554317_7.pdf.

3. Patterson, L. A., and M. W. Doyle. *2020 Aspen-Nicholas Water Forum Water Affordability and Equity Briefing Document.* August 2020. https://nicholasinstitute.duke.edu/publications/2020-aspen-nicholas-water-forum-water-affordability-and-equity-briefing-document.

4. Fedinick, K. P., S. Taylor, and M. Roberts. *Watered Down Justice.* September 2019. https://www.nrdc.org/sites/default/files/watered-down-justice-report.pdf.

5. McGraw., G. "How Do You Fight the Coronavirus without Running Water?" *New York Times*, May 2, 2020. https://www.nytimes.com/2020/05/02/opinion/coronavirus-water.html.

CHAPTER 1

1. Global North and Global South are terms used in lieu of earlier terms that grouped nations in terms of economic development (e.g., developed versus developing). Terms that aim to group nations cannot capture the full complexity and diversity between and within nations and so will always be limiting. The terms *Global North* and *Global South* draw from "an entire history of colonialism, neo-imperialism, and differential economic and social change through which large inequalities in living standard, life expectancy, and access to resources are maintained" (https://Onlineacademiccommunity.Uvic.Ca/Globalsouthpolitics/2018/08/08/Global-South-what-does-it-Mean-and-Why-use-the-Term/). The term Global North includes nations which in the context of the quote were likely former colonizers and for which the term *developed* was used. The term Global South includes nations which were the colonized nations and for which the term *developing* was used.

2. World Health Organization. *Guidelines for Drinking-Water Quality.* 4th ed. WHO Press, 2017.

3. Hanna-Attisha, M. "The Future for Flint's Children." *New York Times,* March 26, 2016. https://www.nytimes.com/2016/03/27/opinion/sunday/the-future-for-flints-children.html.

4. Khokha, S. "California Finally Begins Regulating Cancer-Causing Chemical Found in Drinking Water." *KQED,* July 21, 2017. https://www.kqed.org/science/560344/theres-a-cancer-causing-chemical-in-my-drinking-water-but-california-isnt-regulating-it.

5. McKinley, J. "After Months of Anger in Hoosick Falls, Hearings on Tainted Water Begin." *New York Times,* August 30, 2016. https://www.nytimes.com/2016/08/31/nyregion/hoosick-falls-tainted-water-hearings.html.

6. Briscoe, T. "The Shallowest Great Lake Provides Drinking Water for More People than any Other. Algae Blooms Are Making It Toxic—and It's Getting Worse." *Chicago Tribune,* November 14, 2019. https://www.chicagotribune.com/news/environment/great-lakes/ct-lake-erie-climate-change-algae-blooms-20191114-bjkteorf5vg2hfu3cgqxe2ncru-story.html.

7. Bach, J. "Salem Spending $75 Million to Protect Drinking Water from Toxic Algae." *Statesman Journal,* May 15, 2019. https://www.statesmanjournal.com/story/news/2019/05/15/salem-water-crisis-defence-algae-bloom-cyanotoxins-drinking-water/1120319001/.

8. LaRocco, P., and D. M. Schwartz. "The Grumman Plume: Decades of Deceit." *Newsday,* February 18, 2020. https://projects.newsday.com/long-island/plume-grumman-navy/.

9. Rash, R. "Appalachia's Sacrifice." *New York Times,* November 18, 2016. https://www.nytimes.com/2016/11/19/opinion/appalachias-sacrifice.html.

10. Del Real, J. A. "They Grow the Nation's Food, But They Can't Drink the Water." *New York Times*, May 21, 2019. https://www.nytimes.com/2019/05/21/us/california-central-valley-tainted-water.html?searchResultPosition = 1.

11. London, J., et al. *The Struggle for Water Justice in California's San Joaquin Valley: A Focus on Disadvantaged Unincorporated Communities*. February 2018. https://regionalchange.ucdavis.edu/sites/g/files/dgvnsk986/files/inline-files/The%20Struggle%20for%20Water%20Justice%20FULL%20REPORT.pdf.

12. Dig Deep and US Water Alliance. *Closing the Water Access Gap in the United States: A National Action Plan*. 2019. http://uswateralliance.org/sites/uswateralliance.org/files/Closing%20the%20Water%20Access%20Gap%20in%20the%20United%20States_DIGITAL.pdf.

13. McGraw, G. "For These Americans Clean Water Is a Luxury." *New York Times*, October 20, 2016. https://www.nytimes.com/2016/10/20/opinion/for-these-americans-clean-water-is-a-luxury.html?searchResultPosition = 2.

14. Santos, F. "On Parched Navajo Reservation, 'Water Lady' Brings Liquid Gold." *New York Times*, July 13, 2015. https://www.nytimes.com/2015/07/14/us/on-parched-navajo-reservation-water-lady-brings-liquid-gold.html?searchResultPosition = 1.

15. Bourzac, K. "Monitoring Water Quality After Wildfires." *Chemical & Engineering News* 96, no. 48 (December 2018). https://cen.acs.org/environment/water/Monitoring-water-quality-wildfires/96/i48.

16. Spearing-Bowen, J., and K. Schneider. "Industrial Waste Pollutes America's Drinking Water." Accessed May 25, 2020. https://publicintegrity.org/environment/industrial-waste-pollutes-americas-drinking-water/.

17. Hurdle, J. "NJ Drinking Water Contaminants Increase, Survey Says." Accessed May 25, 2020. https://www.njspotlight.com/2019/10/nj-drinking-water-contaminants-increase-survey-says/.

18. Olson, E. D. "EPA Refuses to Protect Children from Perchlorate-Contaminated Tap Water." Accessed May 25, 2020. https://www.nrdc.org/experts/erik-d-olson/epa-refuses-protect-children-perchlorate-contaminated-tap-water.

CHAPTER 2

1. NASA. "NASA Finds Ancient Organic Material, Mysterious Methane on Mars." Accessed October 6, 2019. https://www.nasa.gov/press-release/nasa-finds-ancient-organic-material-mysterious-methane-on-mars.

2. For an animation describing the greenhouse effect, go to https://www.koshland-science-museum.org/Explore-the-Science/Interactives/what-is-the-Greenhouse-Effect.

3. Diez, A. "Liquid Water on Mars." *Science* 361, no. 6401 (July 25, 2018): 448–49.

4. Lauro, S. E., et al. "Multiple Subglacial Water Bodies Below the South Pole of Mars Unveiled by New MARSIS Data." *Nature Astronomy* 5, no. 1 (2021): 63–70.

5. Mojzsis, S. J., T. M. Harrison, and R. T. Pidgeon. "Oxygen-Isotope Evidence from Ancient Zircons for Liquid Water at the Earth's Surface 4,300 Myr Ago." *Nature* 409, no. 6817 (2001): 178–81.

6. Schopf, J. W., K. Kitajima, M. J. Spicuzza, A. B. Kudryavtsev, and J. W. Valley. "SIMS Analysis of the Oldest Known Assemblage of Microfossils Document Their Taxon-Correlated Carbon Isotope Composition." *Proceedings of the National Academy of Sciences* 115 (2018): 53–58.

7. Service, R. "Seeing the Dawn." *Science* 363, no. 6423 (January 11, 2019): 116–19.

8. Stofan, E. R., et al. "The Lakes of Titan." *Nature* 445, no. 7123 (2007a): 61–64.

9. Overbye, D. "Leagues and Leagues beneath Titan's Methane Seas." *New York Times*, February 23, 2021. https://www.nytimes.com/2021/02/21/science/saturn-titan-moon-exploration.html.

10. Healy, J., R. Fausset, and J. Dobbins. "Cracked Pipes, Frozen Wells, Offline Treatment Plants: A Texan Water Crisis." *New York Times*, February 18, 2021. https://www.nytimes.com/2021/02/19/us/cracked-pipes-frozen-wells-offline-treatment-plants-a-texan-water-crisis.html.

CHAPTER 3

1. UNICEF. "Collecting Water Is Often a Colossal Waste of Time for Women and Girls." August 29, 2016. https://www.unicef.org/media/media_92690.html.

2. WHO/UNICEF Joint Monitoring Programme. "Progress on Drinking Water, Sanitation and Hygiene: 2017 Update and SDG Baselines." July 2017. https://data.unicef.org/resources/progress-drinking-water-sanitation-hygiene-2017-update-sdg-baselines/.

3. Institute for Health Metrics and Evaluation (IHME). "Unsafe Water Source—Level 3 Risk." Accessed March 5, 2021. http://www.healthdata.org/results/gbd_summaries/2019/unsafe-water-source-level-3-risk.

4. Radionuclides, which include radioactive elements like radium and uranium, are also of concern in drinking water. The sources of radionuclides tend to be naturally occurring minerals, although mining activities exacerbate contamination. More details on this group of contaminants can be found at https://www.epa.gov/dwreginfo/radionuclides-rule.

5. Locey, K. J., and J. T. Lennon. "Scaling Laws Predict Global Microbial Diversity." *Proceedings of the National Academy of Sciences* 113, no. 21 (2016): 5970–75.

6. World Health Organization (WHO). "Drinking-Water: Key Facts." Last modified March 21, 2022. https://www.who.int/en/news-room/fact-sheets/detail/drinking-water.

7. World Health Organization Global Health Observatory. "Number of Reported Cases of Cholera." Accessed October 6, 2019. https://www.who.int/data/gho/data/indicators/indicator-details/GHO/number-of-reported-cases-of-cholera.

8. "Cholera in the Americas." *Bulletin of the Pan American Health Organization* 25, no. 3 (1991): 267–73.

9. Domonoske, C. "U.N. Admits Role in Haiti Cholera Outbreak That Has Killed Thousands." *The Two-Way*, NPR, August 18, 2016. https://www.npr.org/sections/thetwo-way/2016/08/18/490468640/u-n-admits-role-in-haiti-cholera-outbreak-that-has-killed-thousands.

10. Domonoske, "U.N. Admits Role."

11. Sengupta, S. "U.N. Apologizes for Role in Haiti's 2010 Cholera Outbreak." *New York Times*, December 1, 2016. https://www.nytimes.com/2016/12/01/world/americas/united-nations-apology-haiti-cholera.html.

12. Pandey, P. K., et al. "Contamination of Water Resources by Pathogenic Bacteria." *AMB Express* 4, no. 51 (2014), https://doi.org/10.1186/s13568-014-0051-x.

13. Choffnes, E. R., and A. Mack. *Global Issues in Water, Sanitation, and Health: Workshop Summary*. National Academies Press, 2009.

14. Corso, P. S., et al. "Costs of Illness in the 1993 Waterborne Cryptosporidium Outbreak, Milwaukee, Wisconsin." *Emerging Infectious Diseases* 9 (2003): 426–31.

15. Widerström, M., et al. "Large Outbreak of Cryptosporidium Hominis Infection Transmitted through the Public Water Supply, Sweden." *Emerging Infectious Diseases* 20 (2014): 581–89.

16. DeSilva, M. B., et al. "Communitywide Cryptosporidiosis Outbreak Associated with a Surface Water-Supplied Municipal Water System—Baker City, Oregon, 2013." *Epidemiology and Infection* 144, no. 2 (2016): 274–84.

17. Centers for Disease Control and Prevention. "Etiology of 928 Drinking Water-Associated Outbreaks, by Year—United States, 1971–2014." https://www.cdc.gov/healthywater/surveillance/pdf/DW_Historical_Figure-H.pdf.

18. Plumer., B. "Frigid Onslaught Stretches Limits of Electric Grids." *New York Times*, February 17, 2021. https://static01.nyt.com/images/2021/02/17/nytfrontpage/scan.pdf.

19. Healy, J., R. Fausset, and J. Dobbins. "Cracked Pipes, Frozen Wells, Offline Treatment Plants: A Texan Water Crisis." *New York Times*, February 19, 2021.

https://www.nytimes.com/2021/02/19/us/cracked-pipes-frozen-wells-offline-treatment-plants-a-texan-water-crisis.html.

20. Firestone, L. "Safe Drinking Water for All." *New York Times*, August 21, 2018.https://www.nytimes.com/2018/08/21/opinion/environment/safe-drinking-water-for-all.html.

21. City of Jackson, Mississippi. "City-Wide Boil Water Advisory Remains in Effect." January 27, 2020. https://www.jacksonms.gov/city-wide-boil-water-advisory-remains-in-effect/.

22. Fentress, E. A., and R. Fausset. "'You Can't Bathe. You Can't Wash.' Water Crisis Hobbles Jackson, Miss., for Weeks." *New York Times*, March 12, 2021. https://www.nytimes.com/2021/03/12/us/jackson-mississippi-water-winter-storm.html.

23. Fentress, "You Can't Bathe."

24. Fentress, "You Can't Bathe."

25. McKinley, J. "Fears about Water Supply Grip Village That Made Teflon Products." *New York Times*, February 29, 2016. https://www.nytimes.com/2016/02/29/nyregion/fears-about-water-supply-grip-village-that-made-teflon-products.html.

26. New York State Department of Health. "Drinking Water Response Activities to Address Local Water Supply Concerns." Last modified September 2021. https://www.health.ny.gov/environmental/investigations/drinkingwaterresponse/.

27. Michigan Radio. "Not Safe to Drink." Accessed October 6, 2019. https://www.michiganradio.org/topic/not-safe-drink#stream/0.

28. Goldbaum, C. "'Tasting Funny for Years': Lead in the Water and a City in Crisis." *New York Times*, August 20, 2019. https://www.nytimes.com/2019/08/20/nyregion/newark-water-crisis.html.

29. Clark, A. *The Poisoned City: Flint's Water and the American Urban Tragedy.* Metropolitan Books, 2018.

CHAPTER 4

1. Mithen, S. *Thirst: Water & Power in the Ancient World.* Phoenix, 2012.

2. Mithen, *Thirst: Water & Power.*

3. Rambeau, C., et al. "Palaeoenvironmental Reconstruction at Beidha, Southern Jordan (C. 18,000–8,500 BP): Implications for Human Occupation during the Natufian and Pre-Pottery Neolithic." In *Water, Life and Civilisation: Climate, Environment and Society in the Jordan Valley,* edited by Emily Black and Steven Mithen, 245–68. Cambridge University Press, 2011.

4. Giosan, L., et al. "Fluvial Landscapes of the Harappan Civilization." *Proceedings of the National Academy of Sciences* 109, no. 26 (May 2012): E1688–94.

5. Turner, B. L., and J. A. Sabloff. "Classic Period Collapse of the Central Maya Lowlands: Insights about Human–Environment Relationships for Sustainability." *Proceedings of the National Academy of Sciences* 109, no. 35 (August 2012): 13908–14.

6. Salzman, J. "Thirst: A Short History of Drinking Water." *Yale Journal of Law & the Humanities* 17 (2006): 94–121.

7. Salzman, J. "Is it Safe to Drink the Water?" *Duke Environmental Law & Policy Forum* 19 (2008): 1–42.

8. Salzman, J. *Drinking Water: A History*. Overlook Duckworth, 2017.

9. Salzman, "Thirst: A Short History."

10. Wolf, A. T. "Indigenous Approaches to Water Conflict Negotiations and Implications for International Water." *International Negotiations* 5 (2000): 357–73.

11. Salzman, *Drinking Water*.

12. US Environmental Protection Agency. "The History of Drinking Water Treatment." Accessed October 6, 2019. https://nepis.epa.gov/Exe/ZyPURL.cgi?Dockey = 200024H9.txt.

13. Salzman, "Thirst: A Short History."

14. Salzman, *Drinking Water*.

15. Wald, C. "The Secret History of Ancient Toilets." *Nature* 533 (May 2016): 456–58.

16. Salzman, *Drinking Water*.

17. Sedlak, D. *Water 4.0: The Past, Present, and Future of the World's Most Vital Resource*. Yale University Press, 2014.

18. Angelakis, A. N., and S. A. Snyder. "Wastewater Treatment and Reuse: Past, Present, and Future." *Water* 7, no. 9 (2015): 4887–95.

19. City of Westminster Archives. "Cholera and the Thames." Accessed October 6, 2019. http://www.choleraandthethames.co.uk.

20. City of Westminster Archives. "Cholera and the Thames."

21. Mann, E. "Story of Cities #14: London's Great Stink Heralds a Wonder of the Industrial World." *The Guardian*, April 4, 2016. http://www.theguardian.com/cities/2016/apr/04/story-cities-14-london-great-stink-river-thames-joseph-bazalgette-sewage-system.

22. Cutler, D. M., and G. Miller. "The Role of Public Health Improvements in Health Advances: The Twentieth-Century United States." *Demography* 42 (February 2005): 1–22.

23. Cutler, "Role of Public Health Improvements."

24. Beach, B., et al. "Typhoid Fever, Water Quality, and Human Capital Development." *Journal of Economic History* 76 (2016): 41–75.

25. Anderson, D. M., et al. "Public Health Efforts and the US Mortality Transition." *The Reporter*, October 2021. National Bureau of Economic Research.

https://www.nber.org/reporter/2021number3/public-health-efforts-and-us-mortality-transition.

26. Beach, B. "Water Infrastructure and Health in U.S. Cities." *Regional Science and Urban Economics* 94 (2022): 103674.

27. Sedlak, *Water 4.0*.

28. Salzman, *Drinking Water*.

29. Sedlak, *Water 4.0*.

30. Cutler, "Role of Public Health Improvements."

31. Sedlak, *Water 4.0*.

32. Sedlak, *Water 4.0*.

33. WHO/UNICEF. "Joint Monitoring Programme Data." Accessed October 6, 2019. https://washdata.org/data/household#!/table?geo0 = region&geo1 = sdg.

34. UNICEF. "Collecting Water Is Often a Colossal Waste of Time for Women and Girls." August 29, 2016. https://www.unicef.org/press-releases/unicef-collecting-water-often-colossal-waste-time-women-and-girls.

35. WaterAid. "Facts and Statistics." Accessed June 20, 2022. https://www.wateraid.org/facts-and-statistics.

36. WaterAid. "Facts and Statistics."

37. Salzman, *Drinking Water*.

38. United Nations. "Millennium Development Goals and Beyond 2015." Accessed October 6, 2019. https://www.un.org/millenniumgoals/.

39. United Nations. "Millennium Declaration." Accessed October 6, 2019. https://www.un.org/esa/devagenda/millennium.html.

40. United Nations. "MDG Goal 7: Ensure Environmental Sustainability." Accessed October 6, 2019. https://sustainabledevelopment.un.org/post2015/transformingourworld.

41. United Nations. "Transforming Our World: The 2030 Agenda for Sustainable Development ." Accessed October 6, 2019. https://sustainabledevelopment.un.org/post2015/transformingourworld.

42. United Nations. "Goal 6: Sustainable Development Knowledge Platform." Accessed October 6, 2019. https://www.un.org/millenniumgoals/environ.shtml.

43. WHO/UNICEF. "Joint Monitoring Programme for Water Supply, Sanitation and Hygiene." Accessed December 23, 2020. https://washdata.org.

44. WHO/UNICEF. "Drinking Water." Accessed December 23, 2020. https://washdata.org/monitoring/drinking-water.

45. WHO/UNICEF. "Water Supply, Sanitation and Hygiene."

46. Legal Defense Fund. *Water/Color: A Study of Race and the Water Affordability Crisis in American Cities*. 2019. https://www.naacpldf.org/our-thinking/issue-report/economic-justice/water-color-a-study-of-race-and-the-water-affordability-crisis-in-americas-cities/.

47. WHO/UNICEF. "Water Supply, Sanitation and Hygiene."

48. WHO/UNICEF. "Water Supply, Sanitation and Hygiene."

49. Dig Deep and US Water Alliance. *Closing the Water Access Gap in the United States: A National Action Plan.* 2019. http://uswateralliance.org/sites/uswateralliance.org/files/Closing%20the%20Water%20Access%20Gap%20in%20the%20United%20States_DIGITAL.pdf.

50. Patterson, L. A., and M. W. Doyle. *2020 Aspen-Nicholas Water Forum: Water Affordability and Equity Briefing Document.* August 2020. https://nicholasinstitute.duke.edu/publications/2020-aspen-nicholas-water-forum-water-affordability-and-equity-briefing-document.

51. Doyle, J. T., et al. "Challenges and Opportunities for Tribal Waters: Addressing Disparities in Safe Public Drinking Water on the Crow Reservation in Montana, USA." *International Journal of Environmental Research and Public Health* 15, no. 4 (2018): 567.

52. Morales, L. "For Some Native Americans, Uranium Contamination Feels Like Discrimination." *All Things Considered*, NPR, November 14, 2017. https://www.npr.org/sections/health-shots/2017/11/14/562856213/for-some-native-americans-uranium-contamination-feels-like-discrimination.

53. Del Real, J. A. "They Grow the Nation's Food, But They Can't Drink the Water." *New York Times,* May 21, 2019. https://www.nytimes.com/2019/05/21/us/california-central-valley-tainted-water.html.

54. Atkin, E. "Is Coal Waste Leaching into America's Drinking Water?" *New Republic*, March 12, 2018. https://newrepublic.com/article/147298/coal-waste-leaching-americas-drinking-water.

55. Zolnikov, T. R., and E. Blodgett-Salafia. "Access to Water Provides Economic Relief through Enhanced Relationships in Kenya." *Journal of Public Health* 39, no. 1 (March 2017): 14–19.

56. US Agency for International Development. *Kenya Integrated Water, Sanitation, and Hygiene.* KIWASH Fact Sheet, February 2021. https://www.usaid.gov/sites/default/files/documents/KIWASH_Fact_Sheet_2021.pdf.

57. Zolnikov, "Access to Water Provides Economic Relief."
58. Zolnikov, "Access to Water Provides Economic Relief."
59. Zolnikov, "Access to Water Provides Economic Relief."
60. Zolnikov, "Access to Water Provides Economic Relief."

61. Venkataraman, B. "Access to Safe Drinking Water: A Paradox in Developed Nations." *Environment: Science and Policy for Sustainable Development* 55, no. 4 (2013): 24–34.

CHAPTER 5

1. *New York Times.* "The Drinking Cup Law." January 28, 1912. https://timesmachine.nytimes.com/timesmachine/1912/01/28/100512238.pdf.

2. Board of Health of the State of New Jersey. *Thirty-Sixth Annual Report of the Board of Health of the State of New Jersey, 1912, and Report of the Bureau of Vital Statistics*. Board of Health of the State of New Jersey, 1913.

3. Treasury Department. "Bacteriological Standard for Drinking Water: The Standard Adopted by the Treasury Department for Drinking Water Supplied to the Public by Common Carriers in Interstate Commerce." *Public Health Reports (1896–1970)* 29, no. 45 (1914): 2959–66.

4. Treasury Department. "Bacteriological Standard for Drinking Water."

5. Okun, D. "Historical Overview of Drinking Water Contaminants and Public Water Utilities." In *Identifying Future Drinking Water Contaminants*, chapter 3. Washington, DC: National Academy Press, 1999. https://nap.nationalacademies.org/read/9595/chapter/3.

6. Hueper, W. C. "Cancer Hazards from Natural and Artificial Water Pollutants." Proceedings from Conference on Physiological Aspects of Water Quality. Washington, DC: US Public Health Service, 1960.

7. US Public Health Service. *Public Health Service Drinking Water Standard*, 1962.

8. *Time*. "America's Sewage System and the Price of Optimism." August 1, 1969. https://content.time.com/time/subscriber/article/0,33009,901182,00.html.

9. Gratani, M., et al. "Indigenous Environmental Values as Human Values." *Cogent Social Sciences* 2, no. 1 (2016): 1185811.

10. US EPA. "Summary of the National Environmental Policy Act." Accessed April 2, 2020. https://www.epa.gov/laws-regulations/summary-national-environmental-policy-act.

11. Lewis, J. "The Birth of EPA." *EPA Journal*, November 1985. https://archive.epa.gov/epa/aboutepa/birth-epa.html.

12. Lewis, "Birth of EPA."

13. US EPA. "Clean Water Act Section 502: General Definitions." Accessed April 2, 2020. https://www.epa.gov/cwa-404/clean-water-act-section-502-general-definitions.

14. US EPA. "Clean Water Act Section 502."

15. US EPA. "Summary of the Safe Drinking Water Act." Accessed October 6, 2019. https://www.epa.gov/laws-regulations/summary-safe-drinking-water-act.

16. US EPA. "Safe Drinking Water Act (SDWA)." Accessed October 6, 2019. https://www.epa.gov/sdwa.

17. Kyros, P. N. "Legislative History of the Safe Drinking Water Act." *Journal of the American Water Works Association* 66 (1974): 566–69.

18. US EPA. "National Primary Drinking Water Regulations." Accessed October 6, 2019. https://www.epa.gov/ground-water-and-drinking-water/national-primary-drinking-water-regulations.

19. US EPA, "Drinking Water Regulations."

20. US EPA, "Drinking Water Regulations."

21. US EPA, "Drinking Water Regulations."

22. Safe Drinking Water Foundation. "What Is Chlorination?" Accessed December 28, 2020. https://www.safewater.org/fact-sheets-1/2017/1/23/what-is-chlorination.

23. Centers for Disease Control and Prevention (CDC). "Free Chlorine Testing" Accessed December 28, 2020. https://www.cdc.gov/safewater/chlorine-residual-testing.html.

24. US EPA. "How EPA Regulates Drinking Water Contaminants." Accessed March 5, 2021. https://www.epa.gov/sdwa/how-epa-regulates-drinking-water-contaminants#standards.

25. US EPA, "Drinking Water Regulations."

26. US EPA. "How EPA Regulates Drinking Water Contaminants."

27. Drinking water regulations for lead are complex, as lead can be introduced into the water distribution system after treatment. To lower this risk, drinking water utilities have to follow treatment techniques as dictated by the Lead and Copper Rule. These treatment techniques aim to keep levels of lead below 15 mcg/L. (For more details, see https://www.epa.gov/dwreginfo/lead-and-copper-rule.)

28. World Health Organization. *Guidelines for Drinking-Water Quality.* 4th ed. WHO Press, 2017.

29. Association of State Drinking Water Administrators. "European Union Updates Drinking Water Standards." March 7, 2019. https://www.asdwa.org/2019/03/07/european-union-updates-drinking-water-standards/.

30. General Secretariat of the Council Delegations. *Proposal for a Directive of the European Parliament and of the Council on the Quality of Water Intended for Human Consumption.* Accessed June 2, 2022. https://op.europa.eu/en/publication-detail/-/publication/13def1fc-5711-11ea-8b81-01aa75ed71a1/language-en.

31. US EPA. "Clean Water Act Section 502."

32. US EPA. "Basic Information about Source Water Protection." Accessed June 2, 2022. https://www.epa.gov/sourcewaterprotection/basic-information-about-source-water-protection.

33. US EPA. "Water on Tap: What You Need to Know." Accessed October 6, 2019. https://nepis.epa.gov/Exe/ZyPDF.cgi/P1008ZP0.PDF?Dockey = P1008ZP0.PDF.

34. Centers for Disease Control and Prevention (CDC). "Water Treatment: Community Water Treatment." Accessed October 6, 2019. https://www.cdc.gov/healthywater/drinking/public/water_treatment.html.

35. Safe Drinking Water Foundation. "Conventional Water Treatment: Coagulation and Filtration." Accessed December 28, 2020. https://www.safewater.org/fact-sheets-1/2017/1/23/conventional-water-treatment.

36. Denver Water. "Virtual Tour: Potable Water Treatment." June 22, 2017. https://www.denverwater.org/education/blog/virtual-tour-potable-water-treatment.

37. Filters used in households also help lower hardness. These filters rely on an ion exchange process.

38. The pH measures a solution's acidity or basicity, with a neutral pH being 7, an acidic pH 7. Pure water has a pH of 7.

39. Per the EPA, "A public water system provides water for human consumption through pipes or other constructed conveyances to at least 15 service connections or serves an average of at least 25 people for at least 60 days a year." (US EPA. "Information about Public Water Systems." Accessed October 6, 2019. https://www.epa.gov/dwreginfo/information-about-public-water-systems.)

40. To learn about ways to protect yourself from lead in drinking water, refer to this site from the Centers for Disease Control and Prevention (CDC): https://www.cdc.gov/nceh/lead/prevention/sources/water.htm.

41. US EPA. "25 Years of the Safe Drinking Water Act: History and Trends." Accessed May 27, 2022. https://www.hsdl.org/?view&did=449348.

42. Weinmeyer, R., A. Norling, M. Kawarski, and E. Higgins. "The Safe Drinking Water Act of 1974 and its Role in Providing Access to Safe Drinking Water in the United States." *AMA Journal of Ethics* 19, no. 10 (2017): 1018–1026.

43. US EPA. "Analyze Trends: EPA/State Drinking Water Dashboard." Accessed April 23, 2020. https://echo.epa.gov/trends/comparative-maps-dashboards/drinking-water-dashboard?view = activity&state = National&yearview = FY&criteria = adv&pwstype = Community%20Water%20System&watersrc = All.

44. US EPA. "Drinking Water Dashboard Help." Accessed April 23, 2020. https://echo.epa.gov/help/drinking-water-dashboard-help.

45. US EPA. "Drinking Water Dashboard Help."

46. Balazs, C., et al. "Social Disparities in Nitrate-Contaminated Drinking Water in California's San Joaquin Valley." *Environmental Health Perspectives* 119, no. 9 (September 2011): 1272–78.

47. Stillo, F., and J. MacDonald Gibson. "Exposure to Contaminated Drinking Water and Health Disparities in North Carolina." *American Journal of Public Health* 107, no. 1 (January 2017): 180–85.

48. Fedinick, K. P., S. Taylor, and M. Roberts. *Watered Down Justice*. Natural Resources Defense Council, September 2019. https://www.nrdc.org/sites/default/files/watered-down-justice-report.pdf?utm_source=tw&utm_medium=twee&utm_campaign=DrinkWater.

49. Patterson, L. A., and M. W. Doyle. *2020 Aspen-Nicholas Water Forum: Water Affordability and Equity Briefing Document*. August 2020. https://nicholasinstitute.duke.edu/publications/2020-aspen-nicholas-water-forum-water-affordability-and-equity-briefing-document.

50. Legal Defense Fund. *Water/Color: A Study of Race and the Water Affordability Crisis in American Cities*. 2019. https://www.naacpldf.org/our-thinking/issue

-report/economic-justice/water-color-a-study-of-race-and-the-water-affordability-crisis-in-americas-cities/.

51. Allaire, M., H. Wu, and U. Lall. "National Trends in Drinking Water Quality Violations." *Proceedings of the National Academy of Sciences* 115, no. 9 (February 2018): 2078–83.

52. Allaire, "National Trends."

53. Fedinick, *Watered Down Justice*.

54. Patterson, *2020 Aspen-Nicholas Water Forum*.

55. Patterson, *2020 Aspen-Nicholas Water Forum*.

56. Fedinick, *Watered Down Justice*.

57. Patterson, *2020 Aspen-Nicholas Water Forum*.

58. Kienzler, A., et al. "Regulatory Assessment of Chemical Mixtures: Requirements, Current Approaches and Future Perspectives." *Regulatory Toxicology and Pharmacology* 80 (October 2016): 321–34.

59. World Health Organization (WHO). *Chemical Mixtures in Source Water and Drinking-Water*. 2017. https://apps.who.int/iris/bitstream/handle/10665/255543/9789241512374-eng.pdf;jsessionid = 91183D7DFB1E35B2494825F4960BD2EE?sequence = 1.

60. Escher, B. I., H. M. Stapleton, and E. L. Schymanski. "Tracking Complex Mixtures of Chemicals in Our Changing Environment." *Science* 367, no. 6476 (January 2020): 388–92.

61. Kienzler, A. "Regulatory Assessment of Chemical Mixtures."

62. WHO. *Chemical Mixtures*.

63. US EPA. "Basic Information on the CCL and Regulatory Determination." Accessed October 6, 2019. https://www.epa.gov/ccl/basic-information-ccl-and-regulatory-determination.

64. US EPA. "Basic Information on the CCL."

65. US EPA. "Basic Information on the CCL."

66. US EPA. "Basic Information on the CCL."

67. Roberson, J. A. "Drinking Water: If We Don't Learn from History We Are Bound to Repeat It." *Journal of American Water Works Association* 111, no. 3 (2019): 18–24.

68. US EPA. "Stronger Protections from Lead in Drinking Water: Next Steps for the Lead and Copper Rule." Accessed August 11, 2022. https://www.epa.gov/system/files/documents/2021-12/lcrr-review-fact-sheet_0.pdf

69. US EPA. "Third Unregulated Contaminant Monitoring Rule." Accessed October 6, 2019. https://www.epa.gov/dwucmr/third-unregulated-contaminant-monitoring-rule.

70. Hofste, R. W., P. Reig, and L. Schleifer. "17 Countries, Home to One-Quarter of the World's Population, Face Extremely High Water Stress." World Resources Institute, August 6, 2019. https://www.wri.org/blog/2019/08

/17-countries-home-one-quarter-world-population-face-extremely-high-water-stress.

71. Flavelle, C. "'Toxic Stew' Stirred Up by Disasters Poses Long-Term Danger, New Findings Show." *New York Times,* July 15, 2019. https://www.nytimes.com/2019/07/15/climate/flooding-chemicals-health-research.html.

72. Pierre-Louis, K., N. Popovich, and H. Tabuchi. "Florence's Floodwaters Breach Coal Ash Pond and Imperil Other Toxic Sites." *New York Times,* September 17, 2018. https://www.nytimes.com/interactive/2018/09/13/climate/hurricane-florence-environmental-hazards.html.

73. Bourzac, K. "Monitoring Water Quality After Wildfires." *Chemical and Engineering News* 96, no. 48 (December 2018). https://cen.acs.org/environment/water/Monitoring-water-quality-wildfires/96/i48.

74. Sengupta, S., and W. Cai. "A Quarter of Humanity Faces Looming Water Crises." *New York Times,* August 6, 2019. https://www.nytimes.com/interactive/2019/08/06/climate/world-water-stress.html.

75. Union of Concerned Scientists. "Water and Climate Change." June 24, 2010. https://www.ucsusa.org/global-warming/science-and-impacts/impacts/water-and-climate-change.html.

76. Melilo, J. M., T. C. Richmond, and G. W. Yohe. *Climate Change Impacts in the United States: The Third National Climate Assessment.* US Global Change Research Program, 2014.

77. American Society of Civil Engineers. "Drinking Water: 2021 Report Card." Accessed November 25, 2021. https://infrastructurereportcard.org/cat-item/drinking-water/.

78. American Society of Civil Engineers, "2021 Report Card."

79. American Society of Civil Engineers. *Failure to Act: Economic Impacts of Status Quo Investment Across Infrastructure Systems.* 2021. https://infrastructurereportcard.org/the-impact/failure-to-act-report/.

80. Patterson, L., and M. Doyle. *Ensuring Water Quality: Innovating on the Clean Water and Safe Drinking Water Acts for the 21st Century: A Report from the 2019 Aspen-Nicholas Water Forum.* 2019. https://nicholasinstitute.duke.edu/sites/default/files/publications/ensuring-water-quality-aspen-nicholas-water-forum-2019.pdf.

81. US Water Alliance. *An Equitable Water Future: A National Briefing Paper.* 2017. http://uswateralliance.org/sites/uswateralliance.org/files/publications/uswa_waterequity_FINAL.pdf.

CHAPTER 6

1. Escher, B. I., H. M. Stapleton, and E. L. Schymanski. "Tracking Complex Mixtures of Chemicals in our Changing Environment." *Science* 367, no. 6476 (January 2020): 388–92.

2. Sen, A., et al. "Multigenerational Epigenetic Inheritance in Humans: DNA Methylation Changes Associated with Maternal Exposure to Lead Can Be Transmitted to the Grandchildren." *Scientific Reports* 5 (2015): 14466.

3. Nriagu, J. O. "Occupational Exposure to Lead in Ancient Times." *Science of the Total Environment* 31, no. 2 (November 1983): 105–16.

4. Rabin, R. "The Lead Industry and Lead Water Pipes: 'A Modest Campaign'." *American Journal of Public Health* 98, no. 9 (September 2008): 1584–92.

5. Rabin, "The Lead Industry."

6. Negus, S. S. "The Physiological Aspects of Mineral Salts in Public Water Supplies." *Journal of the American Water Works Association* 30, no. 2 (February 1938): 242–64.

7. World Health Organization (WHO). "Lead Poisoning and Health." Accessed May 25, 2020. https://www.who.int/news-room/fact-sheets/detail/lead-poisoning-and-health.

8. US EPA. "Lead Ban: Preventing the Use of Lead in Public Water Systems and Plumbing Used for Drinking Water." 1989.

9. SciLine. "Lead in U.S. Drinking Water." American Association for the Advancement of Science, February 22, 2019. https://www.sciline.org/evidence-blog/lead-drinking-water.

10. United Utilities. "Lead Pipes." Accessed April 3, 2020. https://www.unitedutilities.com/lead-pipes.

11. Alberta WaterPortal Society. "Do You Feel Mis-LEAD by Your Water? Facts about Lead in Tap Water." January 21, 2018. https://albertawater.com/do-you-feel-mis-lead-by-your-water/.

12. SciLine, "Lead in U.S. Drinking Water."

13. US EPA. "Lead and Copper Rule." Accessed May 25, 2020. https://www.epa.gov/dwreginfo/lead-and-copper-rule.

14. Keeler, B. L., et al. "Advancing Water Equity Demands New Approaches to Sustainability Science." *One Earth* 2, no. 3 (March 2020): 211–13.

15. Nakamura, D. "Water in D.C. Exceeds EPA Lead Limit." *Washington Post*, January 31, 2004. https://www.washingtonpost.com/archive/politics/2004/01/31/water-in-dc-exceeds-epa-lead-limit/1e54ff9b-a393-4f0a-a2dd-7e8ceedd1e91/.

16. Edwards, M., and A. Dudi. "Role of Chlorine and Chloramine in Corrosion of Lead-Bearing Plumbing Materials." *Journal of the American Water Works Association* 96, no. 10 (October 2004): 69–81.

17. Clark, A. *The Poisoned City: Flint's Water and the American Urban Tragedy*. Metropolitan Books, 2018.

18. Venkataraman, B. "The Paradox of Water and the Flint Crisis." *Environment: Science and Policy for Sustainable Development* 60, no. 1 (2018): 4–17.

19. Pieper, K. J., et al. "Evaluating Water Lead Levels During the Flint Water Crisis." *Environmental Science & Technology* 52, no. 15 (2018): 8124–32.

20. Hanna-Attisha, M., et al. "Elevated Blood Lead Levels in Children Associated with the Flint Drinking Water Crisis: A Spatial Analysis of Risk and Public Health Response." *American Journal of Public Health* 106, no. 2 (2016): 283–90.

21. Venkataraman, "Paradox of Water."

22. Torrice, M. "How Lead Ended Up in Flint's Tap Water." *Chemical and Engineering News* 94, no. 7 (February 2016): 26–29. https://cen.acs.org/articles/94/i7/Lead-Ended-Flints-Tap-Water.html.

23. Torrice, "Flint's Tap Water."

24. Torrice, "Flint's Tap Water."

25. Michigan Civil Rights Commission. *The Flint Water Crisis: Systemic Racism through the Lens of Flint*. Report of the Michigan Civil Rights Commission, February 17, 2017. https://www.michigan.gov/documents/mdcr/VFlintCrisisRep-F-Edited3-13-17_554317_7.pdf.

26. Clark, *Poisoned City*.

27. Legal Defense Fund. *Water/Color: A Study of Race and the Water Affordability Crisis in American Cities*. 2019. https://www.naacpldf.org/our-thinking/issue-report/economic-justice/water-color-a-study-of-race-and-the-water-affordability-crisis-in-americas-cities/.

28. Clark, *Poisoned City*.

29. Fedinick, K. P., S. Taylor, and M. Roberts. *Watered Down Justice*. September 2019. https://www.nrdc.org/sites/default/files/watered-down-justice-report.pdf.

30. Patterson, L. A., and M. W. Doyle. *2020 Aspen-Nicholas Water Forum Water Affordability and Equity Briefing Document*. August 2020. https://nicholasinstitute.duke.edu/publications/2020-aspen-nicholas-water-forum-water-affordability-and-equity-briefing-document.

31. US Census Bureau. "Quick Facts: Flint, Michigan." Accessed December 28, 2020. https://www.census.gov/quickfacts/fact/table/flintcitymichigan,US/PST045219.

32. Leyden, L. A. "Water Crisis in Newark Brings New Worries." *New York Times*, December 3, 2018. https://www.nytimes.com/2018/12/03/nyregion/newark-drinking-water-lead.html.

33. Corasaniti, N., C. Kilgannon, and J. Schwartz. "Tainted Water, Ignored Warnings and a Boss with a Criminal Past." *New York Times*, August 24, 2019. https://www.nytimes.com/2019/08/24/nyregion/newark-lead-water-crisis.html.

34. Goldbaum, C. "'Tasting Funny for Years:' Lead in the Water and a City in Crisis." *New York Times*, August 20, 2019. https://www.nytimes.com/2019/08/20/nyregion/newark-water-crisis.html.

35. Olson, E. D., and K. P. Fedinick. "What's in Your Water? Flint and Beyond." Natural Resources Defense Council, June 28, 2016. https://www.nrdc.org/resources/whats-your-water-flint-and-beyond.

36. US EPA. "Stronger Protections from Lead in Drinking Water: Next Steps for the Lead and Copper Rule." Accessed August 11, 2022. https://www.epa.gov/system/files/documents/2021-12/lcrr-review-fact-sheet_0.pdf

37. Pelley, J. "Treatment for Lead in Drinking Water Is Evolving. Will the U.S. EPA Catch Up?" *Chemical and Engineering News,* November 21, 2018. https://cen.acs.org/environment/water/Treatment-lead-drinking-water-evolving/96/i47.

38. Natural Resources Defense Council. "Concerned Pastors for Social Action v. Khouri." Last modified December 1, 2020. https://www.nrdc.org/court-battles/concerned-pastors-social-action-v-khouri-16-cv-10277-ed-mich.

39. Corasaniti, N. "Newark Water Crisis: Racing to Replace Lead Pipes in Under 3 Years." *New York Times,* August 26, 2019. https://www.nytimes.com/2019/08/26/nyregion/newark-lead-water-pipes.html.

40. Environmental Defense Fund. "State Efforts to Support LSL Replacement." Accessed April 3, 2020. https://www.edf.org/health/state-efforts-support-lsl-replacement.

41. Environmental Defense Fund. "Community and Utility Efforts to Replace Lead Service Lines." Accessed April 3, 2020. https://www.edf.org/health/recognizing-community-efforts-replace-lsl.

42. The White House. "Fact Sheet: The American Jobs Plan." March 31, 2021. https://www.whitehouse.gov/briefing-room/statements-releases/2021/03/31/fact-sheet-the-american-jobs-plan/.

43. Campbell, S., and D. Wesssel. "What Would It Cost to Replace All the Nation's Lead Water Pipes?" Brookings, May 13, 2021. https://www.brookings.edu/blog/up-front/2021/05/13/what-would-it-cost-to-replace-all-the-nations-lead-water-pipes/.

44. Haber is a controversial figure. While his work did help feed a growing population, he is also referred to as the "father of chemical warfare."

45. Erisman, J. W., et al. "How a Century of Ammonia Synthesis Changed the World." *Nature Geoscience* 1 (October 2008): 636–39.

46. Erisman, "Ammonia Synthesis."

47. Ritchie, H. "How Many People Does Synthetic Fertilizer Feed?" Our World in Data, November 7, 2017. https://ourworldindata.org/how-many-people-does-synthetic-fertilizer-feed.

48. Erisman, "Ammonia Synthesis."

49. Agency for Toxic Substances and Disease Registry. "What Are the Health Effects from Exposure to Nitrates and Nitrites?" Accessed April 3, 2020. https://www.atsdr.cdc.gov/csem/nitrate-nitrite/health_effects.html.

50. Ward, M. H., et al. "Drinking Water Nitrate and Human Health: An Updated Review." *International Journal of Environmental Research and Public Health* 15, no. 7 (2018): 1557.

51. Ward, "Drinking Water Nitrate."

52. Mateo-Sagasta, J., S. M. Zadeh, and H. Turral. *More People, More Food, Worse Water? A Global Review of Water Pollution from Agriculture.* Food and Agriculture Organization of the United Nations, June 20, 2018. https://reliefweb.int/report/world/more-people-more-food-worse-water-global-review-water-pollution-agriculture.

53. Del Real, J. A. "They Grow the Nation's Food, But They Can't Drink the Water." *New York Times,* May 21, 2019. https://www.nytimes.com/2019/05/21/us/california-central-valley-tainted-water.html.

54. Schaider, L. A., et al. "Environmental Justice and Drinking Water Quality: Are There Socioeconomic Disparities in Nitrate Levels in U.S. Drinking Water?" *Environmental Health* 18, no. 1 (2019): 3.

55. Schechinger, A. W. "America's Nitrate Habit Is Costly and Dangerous." EWG, October 2, 2018. https://www.ewg.org/research/nitratecost/.

56. Rezvani, F., et al. "Nitrate Removal from Drinking Water with a Focus on Biological Methods: A Review." *Environmental Science and Pollution Research* 26 (2019): 1124–41.

57. CFCs are also responsible for depleting levels of ozone in the stratosphere, forming the so-called "ozone hole."

58. Rich, N. "The Lawyer Who Became DuPont's Worst Nightmare." *New York Times,* January 4, 2016. https://www.nytimes.com/2016/01/10/magazine/the-lawyer-who-became-duponts-worst-nightmare.html.

59. Rich, "Lawyer."

60. Steenland, K., T. Fletcher, and D. A. Savitz. "Epidemiologic Evidence on the Health Effects of Perfluorooctanoic Acid (PFOA)." *Environmental Health Perspectives* 118, no. 8 (2010): 1100–1108.

61. Fletcher, T., D. Savitz, and K. Steenland. "C8 Science Panel." Last modified January 22, 2020. http://www.c8sciencepanel.org.

62. UN Stockholm Convention. "The New POPs under the Stockholm Convention." Accessed May 27, 2022. http://chm.pops.int/TheConvention/ThePOPs/TheNewPOPs/tabid/2511/Default.aspx.

63. US EPA. "Third Unregulated Contaminant Monitoring Rule." Accessed October 6, 2019. https://www.epa.gov/dwucmr/third-unregulated-contaminant-monitoring-rule.

64. US EPA. *Fact Sheet: PFOA & PFOS Drinking Water Health Advisories.* November 2016. https://www.epa.gov/sites/default/files/2016-06/documents/drinkingwaterhealthadvisories_pfoa_pfos_updated_5.31.16.pdf.

65. Environmental Working Group. "Mapping a Contamination Crisis." June 8, 2017. https://www.ewg.org/research/mapping-contamination-crisis.

66. Reade, A., and T. Quinn. *Scientific and Policy Assessment for Addressing Per- and Polyfluoroalkyl Substances (PFAS) in Drinking Water*. Natural Resources Defense Council, March 15, 2019. https://www.nrdc.org/sites/default/files/assessment-for-addressing-pfas-chemicals-in-michigan-drinking-water.pdf.

67. Andrews, D. Q., and O. V. Naidenko. "Population-Wide Exposure to Per- and Polyfluoroalkyl Substances from Drinking Water in the United States." *Environmental Science & Technology Letters* 7, no. 12 (2020): 931–36.

68. Hogue, C. "A Guide to the PFAS Found in Our Environment." *Chemical and Engineering News*. Accessed October 9, 2019. https://cen.acs.org/sections/pfas.html.

69. Australian Government Department of Health. "Per- and Poly-fluoroalkyl Substances (PFAS)." Last modified May 10, 2022. https://www1.health.gov.au/internet/main/publishing.nsf/Content/ohp-pfas.htm.

70. IPEN. *PFAS Pollution across the Middle East and Asia*. April 2019. https://ipen.org/sites/default/files/documents/pfas_pollution_across_the_middle_east_and_asia.pdf.

71. Goldenman, G., et al. *The Cost of Inaction: A Socioeconomic Analysis of Environmental and Health Impacts Linked to Exposure to PFAS*. Nordic Council of Ministers, 2019. https://norden.diva-portal.org/smash/get/diva2:1295959/FULLTEXT01.pdf.

72. Blum, A. et al. "The Madrid Statement on Poly- and Perfluoroalkyl Substances (PFASs)." *Environmental Health Perspectives* 123, no. 5 (2015): A107–11.

73. IPEN. "At UN Meeting, Governments Agree to a Global Ban on PFOA—a Toxic Water Pollutant." May 3, 2019. https://ipen.org/news/un-meeting-governments-agree-global-ban-pfoa---toxic-water-pollutant.

74. Hopkins, Z. R., et al. "Recently Detected Drinking Water Contaminants: GenX and Other Per- and Polyfluoroalkyl Ether Acids." *Journal of American Water Works Association* 110, no. 7 (July 2018): 13–28.

75. Conley, J. M., et al. "Adverse Maternal, Fetal, and Postnatal Effects of Hexafluoropropylene Oxide Dimer Acid (GenX) from Oral Gestational Exposure in Sprague-Dawley Rats." *Environmental Health Perspectives* 127, no. 3 (2019).

76. Hagerty, V. "Toxin Taints Cape Fear River South of Fayetteville Works Plant." *Fayetteville Observer*, June 7, 2017. https://www.fayobserver.com/story/news/2017/06/07/toxin-taints-cape-fear-river-south-of-fayetteville-works/20685899007/.

77. Sun, M., et al. "Legacy and Emerging Perfluoroalkyl Substances Are Important Drinking Water Contaminants in the Cape Fear River Watershed of North Carolina." *Environmental Science & Technology Letters* 3, no. 12 (2016): 415–19.

78. Hogue, C. "Environmental Groups Seek Order from North Carolina to Stop Chemours Pollution." *Chemical and Engineering News* 96 (2018). https://cen.acs.org/policy/litigation/Environmental-groups-seek-order-North/96/i20.

79. "North Carolina PFAS Testing Network." Accessed October 9, 2019. https://ncpfastnetwork.com.

80. Lerner, S. "Citizen Groups Will Sue DuPont and Chemours for Contaminating Drinking Water in North Carolina." *The Intercept*, August 8 2017. https://theintercept.com/2017/08/08/citizen-groups-will-sue-dupont-and-chemours-for-contaminating-drinking-water-in-north-carolina/.

81. Hogue, C. "The Hunt Is on for GenX Chemicals in People." *Chemical and Engineering News* 97, no. 15 (2019): 23.

82. Tiemann, M., and E. H. Humphreys. *Regulating Drinking Water Contaminants: EPA PFAS Action*. Congressional Research Service, February 26, 2020. https://sgp.fas.org/crs/misc/IF11219.pdf.

83. US EPA. "EPA Announces Proposed Decision to Regulate PFOA and PFOS in Drinking Water." Accessed May 26, 2020. https://www.epa.gov/newsreleases/epa-announces-proposed-decision-regulate-pfoa-and-pfos-drinking-water.

84. Safer States. "PFAS." Accessed October 9, 2019. http://www.saferstates.com/toxic-chemicals/pfas/.

85. Talpos, S. "They Persisted." *Science* 364, no. 6441 (2019): 622–26.

86. US EPA. "EPA Takes Action to Address PFAS in Drinking Water." Accessed March 1, 2021. https://www.epa.gov/newsreleases/epa-takes-action-address-pfas-drinking-water.

87. Hogue, C. "Checking Drinking Water for PFAS and Lithium." *Chemical and Engineering News* 99, no. 7 (2021): 16.

88. US EPA. "Announcement of Final Regulatory Determinations for Contaminants on the Fourth Drinking Water Contaminant Candidate List." Accessed March 12, 2021. https://www.federalregister.gov/documents/2021/03/03/2021-04184/announcement-of-final-regulatory-determinations-for-contaminants-on-the-fourth-drinking-water.

89. US EPA. "Final Regulatory Determinations."

90. US EPA. "PFAS Strategic Roadmap: EPA's Commitments to Action 2021–2024." Accessed December 12, 2021. https://www.epa.gov/pfas/pfas-strategic-roadmap-epas-commitments-action-2021-2024.

91. Hogue, C. "US Drinking Water to Be Tested for 29 PFAS." *Chemical and Engineering News*, December 22, 2021. https://cen.acs.org/environment/persistent-pollutants/US-drinking-water-tested-29/99/web/2021/12.

CHAPTER 7

1. Snider, A. "What Broke the Safe Drinking Water Act?" *Politico*, May 10, 2017. https://www.politico.com/agenda/story/2017/05/10/safe-drinking-water-perchlorate-000434/.

2. Natural Resources Defense Council. "EPA Blows Up Court Order Over Rocket Fuel Contaminant in Drinking Water." June 18, 2020. https://www.nrdc.org/media/2020/200618-0.

3. US EPA. "Perchlorate in Drinking Water: Frequently Asked Questions." Accessed November 14, 2021. https://www.epa.gov/sdwa/perchlorate-drinking-water-frequent-questions.

4. US EPA. "EPA Issues Final Action for Perchlorate in Drinking Water." Accessed July 13, 2020. https://www.epa.gov/newsreleases/epa-issues-final-action-perchlorate-drinking-water.

5. Olson, E. D. "NRDC Sues to Protect Kids from Perchlorate in Tap Water." Natural Resources Defense Council, September 3, 2020. https://www.nrdc.org/experts/erik-d-olson/nrdc-sues-protect-kids-perchlorate-tap-water.

6. Erickson, B. E. "How Many Chemicals Are in Use Today? EPA Struggles to Keep Its Chemical Inventory Up to Date." *Chemical and Engineering News* 95, no. 9 (2017): 23–24.

7. Johnson, A. C., et al. "Learning from the Past and Considering the Future of Chemicals in the Environment." *Science* 367, no. 6476 (2020): 384–87.

8. Snider, "Safe Drinking Water Act."

9. Siegel, S. M. *Troubled Water: What's Wrong with What We Drink*. Thomas Dunne Books, 2019.

10. O'Riordan, T., and J. Cameron, eds. *Interpreting the Precautionary Principle*. Earthscan, 1994.

11. Martuzzi, M., and J. A. Tickner, eds. *The Precautionary Principle: Protecting Public Health, the Environment and the Future of Our Children*. World Health Organization, 2004.

12. Percival, R. V. "Who's Afraid of the Precautionary Principle?" *Pace Environmental Law Review* 23 (2005–2006): 21–81.

13. Persson, E. "What Are the Core Ideas Behind the Precautionary Principle?" *Science of the Total Environment* 557–558 (July 2016): 134–41.

14. United Nations. *Rio Declaration on Environment and Development*. Report of the United Nations Conference on Environment and Development, 1992.

15. Martuzzi, *The Precautionary Principle*.

16. Percival, "Who's Afraid."

17. Persson, "What Are the Core Ideas."

18. Boyer-Kassem, T. "Philosophy and the Precautionary Principle: Science, Evidence, and Environmental Policy." *Ethics, Policy and Environment* 22 (2019): 103–5.

19. Science for Environment Policy. *The Precautionary Principle: Decision Making Under Uncertainty*. Future Brief 18, September 2017.

20. Pinto-Bazurco, J. F. *The Precautionary Principle*. International Institute for Sustainable Development, Brief #4, October 2020.

21. Fischer, A. J., and G. Ghelardi. "The Precautionary Principle, Evidence-Based Medicine, and Decision Theory in Public Health Evaluation." *Frontiers in Public Health* 4 (2016): 107.

22. Fischer, "The Precautionary Principle."

23. European Chemicals Agency. "Understanding REACH." Accessed February 18, 2020. https://echa.europa.eu/regulations/reach/understanding-reach.

24. US EPA. "Summary of the Toxic Substances Control Act." Accessed March 9, 2020. https://www.epa.gov/laws-regulations/summary-toxic-substances-control-act.

25. Environmental Working Group. "Chemical Policy." Accessed February 18, 2020. https://www.ewg.org/key-issues/toxics/chemical-policy.

26. Denison, R. A. "A Primer on the New Toxic Substances Control Act (TSCA) and What Led to It." Environmental Defense Fund, April 2017. https://www.edf.org/sites/default/files/denison-primer-on-lautenberg-act.pdf.

27. Erickson, B. "Groups Urge US EPA to Strengthen PFAS Rule." *Chemical and Engineering News* 98, no. 16 (2020): 15.

28. European Chemicals Agency. "Perfluoroalkyl chemicals (PFAS)." Accessed February 8, 2020. https://echa.europa.eu/en/hot-topics/perfluoroalkyl-chemicals-pfas.

29. Hogue, C. "EU Countries Planning PFAS Restrictions." *Chemical and Engineering News* 98, no. 20 (2020): 15.

CHAPTER 8

1. Porter, K. S. "Fixing Our Drinking Water: From Field and Forest to Faucet." *Pace Environmental Law Review* 23 (Summer 2006): 389–422.

2. US EPA. "Source Water Protection Program." Accessed February 25, 2020. https://www.epa.gov/sourcewaterprotection/basic-information-about-source-water-protection.

3. Pollans, M. J. "Drinking Water Protection and Agricultural Exceptionalism." *Ohio State Law Journal* 77 (2016): 1195–1260.

4. Associated Press. "W. Va. Gov Declares Emergency After Chemical Spill." *Bluefield Daily Telegraph*, January 9, 2014. https://www.bdtonline.com/archives/w-va-gov-declares-emergency-after-chemical-spill/article_4d8264d6-9b22-5762-80d7-69040deec947.html.

5. Gabriel, T. "Thousands without Water After Spill in West Virginia." *New York Times*, January 10, 2014. https://www.nytimes.com/2014/01/11/us/west-virginia-chemical-spill.html.

6. Jonsson, P. "West Virginia Chemical Spill: Does It Threaten Clean Water Gains." *Christian Science Monitor*, January 11, 2014. https://www.csmonitor

.com/Environment/2014/0111/West-Virginia-chemical-spill-Does-it-threaten-clean-water-gains.

7. US EPA. *Hydraulic Fracturing for Oil and Gas: Impacts from the Hydraulic Fracturing Water Cycle on Drinking Water Resources in the United States: Executive Summary*. 2016.

8. Pollans, "Drinking Water Protection."

9. McGregor, D. "Traditional Knowledge: Considerations for Protecting Water in Ontario." *International Indigenous Policy Journal* 3, no. 3 (2012).

10. McGregor, D. "Traditional Knowledge and Water Governance: The Ethic of Responsibility." *AlterNative: An International Journal of Indigenous Peoples* 10 (2014): 492–507.

11. National Oceanic and Atmospheric Administration (NOAA). "What Is a Watershed?" Accessed December 12, 2021. https://oceanservice.noaa.gov/facts/watershed.html.

12. Millennium Ecosystem Assessment Board. *Ecosystems and Human Well-being: Wetlands and Water Synthesis*. 2005. https://wedocs.unep.org/20.500.11822/8735.

13. Sindelar, M. "Soils Clean and Capture Water." April 2015. https://www.soils.org/files/sssa/iys/april-soils-overview.pdf.

14. Salt, D. E., R. D. Smith, and I. Raskin. "Phytoremediation." *Annual Reviews of Plant Physiology and Plant Molecular Biology* 49 (1998): 643–48.

15. Ozment, S., et al. *Protecting Drinking Water at the Source: Lessons from United States Watershed Investment Programs*. World Resources Institute, 2016.

16. Ozment, *Protecting Drinking Water at the Source*.

17. Gartner, T., et al. *Natural Infrastructure: Investing in Forested Landscapes for Source Water Protection in the United States*. World Resources Institute, 2013.

18. Smith, L., et al., eds. *Catchment and River Basin Management: Integrating Science and Governance*. Routledge, 2015.

19. Dudley, N., and S. Stolton. *Running Pure: The Importance of Forest Protected Areas for Drinking Water*. World Bank/WWF Alliance for Forest Conservation and Sustainable Use, 2003.

20. Finnegan, M. C. "New York City's Watershed Agreement: A Lesson in Sharing Responsibility." *Pace Environmental Law Review* 14 (1997): 577–644.

21. Soll, D. *Empire of Water: An Environmental and Political History of the New York City Supply*. Cornell University Press, 2013.

22. Soll, *Empire of Water*.

23. Finnegan, "New York City's Watershed Agreement."

24. Soll, *Empire of Water*.

25. Finnegan, "New York City's Watershed Agreement."

26. Soll, *Empire of Water*.

27. Finnegan, "New York City's Watershed Agreement."

28. US EPA. "Surface Water Treatment Rule." Accessed February 21, 2020. https://www.epa.gov/dwreginfo/surface-water-treatment-rules.

29. Hanlon, J. W. "Watershed Protection to Secure Ecosystem Services: The New York City Watershed Governance Arrangement." *Case Studies in the Environment* 1, no. 1 (December 2017): 1–6.

30. Hanlon, "Watershed Protection."

31. New York City (NYC) Environmental Protection. *New York City Drinking Water Supply and Quality Report – 2021*. Accessed May 27, 2022. https://www1.nyc.gov/assets/dep/downloads/pdf/water/drinking-water/drinking-water-supply-quality-report/2021-drinking-water-supply-quality-report.pdf.

32. NYC Environmental Protection, *Drinking Water Supply and Quality Report*.

33. New York State Department of Environmental Conservation. "New York City Water Supply." Accessed January 25, 2020. https://www.dec.ny.gov/lands/25599.html.

34. Dunlap, D. "As a Plant Nears Completion, Croton Water Flows Again to New York City." *New York Times*, May 8, 2015. https://www.nytimes.com/2015/05/09/nyregion/croton-water-is-once-again-flowing-to-new-york.html.

35. Mell, I. C. "Green Infrastructure: Concepts and Planning." *FORUM Journal* 8 (2008): 69–80.

36. Natural Resources Defense Council. "Encourage Green Infrastructure." Accessed January 23, 2020. https://www.nrdc.org/issues/encourage-green-infrastructure.

37. Canning, J. F., and A. S. Stillwell. "Nutrient Reduction in Agricultural Green Infrastructure: An Analysis of the Raccoon River Watershed." *Water* 10, no. 6 (2018): 749.

38. Sun Chan, F. K., et al. "'Sponge City' in China: A Breakthrough in Planning and Flood Risk Management in the Urban Context." *Land Use Policy* 76 (2018): 772–78.

39. Keller, B. L., et al. "Advancing Water Equity Demands New Approaches to Sustainability Science." *One Earth* 2 (2020): 211–13.

40. Keller, "Advancing Water Equity."

41. Southface. "Green Infrastructure and Resilience Institute." Accessed March 26, 2020. https://www.southface.org/programs/giri/.

42. Klein, M., et al. *Sharing in the Benefits of a Greening City: A Policy Toolkit in Pursuit of Economic, Environmental and Racial Justice*. Institute on the Environment, University of Minnesota, 2020.

43. Sucher, K. "A New Water Story: In Conversation with Sandra Postel." Island Press, October 10, 2017. https://islandpress.org/blog/new-water-story-conversation-sandra-postel.

44. Sengupta, S., C. Einhorn, and M. Andreoni. "There's a Global Plan to Conserve Nature: Indigenous People Could Lead the Way." *New York Times*, March 11,

2021. https://www.nytimes.com/2021/03/11/climate/nature-conservation-30-percent.html.

45. Hohner, A. K., et al. "Wildfires Alter Forest Watersheds and Threaten Drinking Water Quality." *Accounts of Chemical Research* 52, no. 5 (2019): 1234–44.

46. Davenport, C. "Trump Removes Pollution Controls on Streams and Wetlands." *New York Times*, January 22, 2020. https://www.nytimes.com/2020/01/22/climate/trump-environment-water.html.

47. Flavelle, C. "E.P.A. Is Letting Cities Dump More Raw Sewage into Rivers for Years to Come." *New York Times*, January 24, 2020. https://www.nytimes.com/2020/01/24/climate/epa-sewage-rivers.html.

CHAPTER 9

1. World Health Organization (WHO). *Potable Reuse: Guidance for Producing Safe Drinking-Water.* 2017. https://apps.who.int/iris/bitstream/handle/10665/258715/9789241512770-eng.pdf?sequence = 1.

2. Angelakis, A. N., et al. "Water Reuse: From Ancient to Modern Times and the Future." *Frontiers in Environmental Science* 6, no. 26 (2018): 1–17.

3. United Nations (UN). *Wastewater: The Untapped Resource.* United Nations World Water Development Report, 2017.

4. UN, *Wastewater*.

5. UN, *Wastewater*.

6. Katsnelson, A. "Urban Stormwater Presents Pollution Challenge." *Chemical and Engineering News* 100 (2022): 13–15.

7. UN, *Wastewater*.

8. WHO, *Potable Reuse*.

9. Gerrity, D., et al. "Potable Reuse Treatment Trains Throughout the World." *Journal of Water Supply: Research and Technology* 62 (2013): 321–38.

10. Gerrity, "Potable Reuse Treatment Trains."

11. National Research Council. *Understanding Water Reuse: Potential for Expanding the Nation's Water Supply through Reuse of Municipal Wastewater.* National Academies Press, 2012. https://www.nap.edu/catalog/13514/understanding-water-reuse-potential-for-expanding-the-nations-water-supply.

12. Metzler, D. F., and H. B. Russelman. "Wastewater Reclamation as a Water Resource." *Journal of American Water Works Association* 60 (1968): 95–102.

13. Hummer, N., and S. Eden. "Potable Reuse of Water." *Arroyo*, 2016.

14. Hummer, "Potable Reuse."

15. WHO, *Potable Reuse*.

16. WHO, *Potable Reuse*.

17. City of San Diego Public Utilities Department. *Pure Water San Diego*. February 3, 2015. https://www.sandiego.gov/sites/default/files/legacy/water/pdf/purewater/2015/faq_purewater.pdf.

18. WHO, *Potable Reuse*.

19. American Water Works Association. *Potable Reuse 101: An Innovative and Sustainable Water Supply Solution*. 2016. https://www.awwa.org/Portals/0/AWWA/ETS/Resources/PotableReuse101.pdf?ver = 2018-12-12-182505-710.

20. City of San Diego. "Pure Water San Diego Virtual Tour." Accessed December 11, 2021. https://www.sandiego.gov/public-utilities/sustainability/pure-water-sd/virtual-tour.

21. As a comparison, the pore size in home-based water filters is on the order of 1 micron.

22. Osmosis is the natural migration of water across a semipermeable membrane from a region where it is at a higher concentration to a region where it is at a lower concentration. For example, if a semipermeable membrane separates saltwater and freshwater solutions, water will flow from the freshwater side to the saltwater side. Reverse osmosis (RO) is the flow of water across a semipermeable membrane against this natural tendency. For water to flow from the saltwater side to the freshwater side, pressure must be applied to overcome the natural osmotic process. To achieve this, an external source of energy pressurizes the water in the saltwater side and forces it to flow across the semipermeable membrane to the freshwater side. The higher the salt concentration of the water, the more energy is needed to pressurize the water to flow from one side to the other. The high salt content of oceans makes desalination by RO an energy-intensive way to produce fresh water.

23. Vandegrift, J., et al. "Overview of Monitoring Techniques for Evaluating Water Quality at Potable Reuse Treatment Facilities." *Journal of American Water Works Association* 111, no. 7 (2019): 12–23.

24. Person, B. M., et al. "Achieving Reliability in Potable Reuse: The Four Rs." *Journal of American Water Works Association* 107 (2015): 48–58.

25. Xiong, J., et al. "The Rejection of Perfluoroalkyl Substances by Nanofiltration and Reverse Osmosis: Influencing Factors and Combination Processes." *Environmental Science Water Research & Technology* 7 (2021): 1928–43.

26. Mastropietro, T. F., et al. "Reverse Osmosis and Nanofiltration Membranes for Highly Efficient PFASs Removal: Overview, Challenges and Future Perspectives." *Dalton Transactions* 50 (2021): 5398–5410.

27. Stoiber, T., et al. *PFAS in Drinking Water: An Emergent Water Quality Threat*. Water Solutions, 2020. https://www.ewg.org/sites/default/files/u352/Stoiber_Evans_WaterSolutions_2020.pdf.

28. Xiong, "Rejection of Perfluoroalkyl Substances."

29. Johnson, A. C., et al. "Identification and Quantification of Microplastics in Potable Water and Their Sources within Water Treatment Works in England and Wales." *Environmental Science and Technology* 54 (2020): 12326–34.

30. Shen, M., et al. "Removal of Microplastics Via Drinking Water Treatment: Current Knowledge and Future Directions." *Chemosphere* 251 (July 2020): 126612.

31. Dalmau-Soler, J., et al. "Microplastics from Headwaters to Tap Water: Occurrence and Removal in a Drinking Water Treatment Plant in Barcelona Metropolitan Area (Catalonia, NE Spain)." *Environmental Science and Pollution Research* 28 (2021): 59462–72.

32. Feld, L., et al. "A Study of Microplastic Particles in Danish Tap Water." *Water* 13, no. 15 (2021): 2097.

33. Stoiber, *PFAS in Drinking Water*.

34. Shen, "Removal of Microplastics."

35. Poerio, T., E. Piacentini, and R. Mazzei. "Membrane Processes for Microplastic Removal." *Molecules* 24, no. 22 (2019): 4148.

36. Tang, K. H. D., and T. Hadibarata. "Microplastics Removal through Water Treatment Plants: Its Feasibility, Efficiency, Future Prospects and Enhancement by Proper Waste Management." *Environmental Challenges* 5 (December 2021): 100264.

37. Kienzler, A., et al. "Regulatory Assessment of Chemical Mixtures: Requirements, Current Approaches and Future Perspectives." *Regulatory Toxicology and Pharmacology* 80 (October 2016): 321–34.

38. Escher, B. I., H. M. Stapleton, and E. L. Schymanski. "Tracking Chemical Mixtures in Our Changing Environments." *Science* 367, no. 6476 (2020): 388–92.

39. Orange County Water District. "GWRS – Ground Water Replenishment System." Accessed January 9, 2020. https://www.ocwd.com/gwrs/.

40. Lefebvre, O. "Beyond NEWater: An Insight into Singapore's Water Reuse Prospects." *Current Opinion in Environmental Science & Health* 2 (April 2018): 26–31.

41. PUB. *Our Water, Our Future*. PUB, Singapore's National Water Agency, June 2016. https://www.pub.gov.sg/Documents/PUBOurWaterOurFuture.pdf.

42. PUB. "NEWater Quality." PUB, Singapore's National Water Agency. Accessed January 9, 2020. https://www.pub.gov.sg/watersupply/waterquality/newater.

43. PUB, "NEWater Quality."

44. Tortajada, C., and P. van Rensburg. "Drink More Recycled Wastewater." *Nature* 577 (2020): 26–28.

45. Hummer, "Potable Reuse."

46. Du Pisani, P., and J. G. Menge. "Direct Potable Reclamation in Windhoek: A Critical Review of the Design Philosophy of New Goreangab Drinking Water Reclamation Plant." *Water Science & Technology: Water Supply* 13 (2013): 214–26.

47. Lahnsteiner, J., P. van Rensburg, and J. Esterhuizen. "Direct Potable Reuse—A Feasible Water Management Option." *Journal of Water Reuse and Desalination* 8 (2018): 14–27.

48. Lewis, E. W., C. Staddon, and J. Sirunda. "Urban Water Management Challenges and Achievements in Windhoek, Namibia." *Water Practice and Technology* 14, no. 3 (2019): 703–13.

49. Du Pisani, "Direct Potable Reclamation."

50. Du Pisani, "Direct Potable Reclamation."

51. van Rensberg, P. "History of the First Ever Pipe-to-Pipe Direct Potable Reuse Facility in Namibia." YouTube video, August 9, 2018. https://www.youtube.com/watch?v=lv4hHNSlh_Q.

52. Hanna, B. "Wichita Falls Says Goodbye to Potty Water for Now." *Washington Times*, July 28, 2015. https://www.washingtontimes.com/news/2015/jul/28/wichita-falls-says-goodbye-to-potty-water-for-now/.

53. City of San Diego. *Pure Water San Diego Program*. December 2021. https://www.sandiego.gov/sites/default/files/pure_water_main_fact_sheet_1.12.22.pdf.

54. City of San Diego Public Utilities Department. *Pure Water*.

55. City of San Diego Public Utilities Department. *Pure Water*.

56. City of San Diego Public Utilities Department. *Pure Water*.

57. Udasin, S. "In an Arid US West Water Agencies Look to Deliver Purified Wastewater Directly to Customers' Faucets, Despite 'Yuck Factor.'" Ensia, August 3, 2021. https://ensia.com/features/in-an-arid-u-s-west-water-agencies-look-to-delive%E2%80%8Br%E2%80%8B-purified-wastewater-directly-to-customers-faucets%E2%80%8B-despite-yuck-factor/.

58. Oglesby, I., and B. Delgado. "California Coastal Commission Must Halt Unjust, Destructive Monterey Desalination Plant." *Sacramento Bee*, September 16, 2020. https://www.sacbee.com/opinion/op-ed/article245764250.html.

59. Xia, R. "Water Company Withdraws Desalination Proposal as Battle Over Environmental Justice Heats Up." *Los Angeles Times*, September 16, 2020. https://www.latimes.com/california/story/2020-09-16/monterey-bay-desalination-plant-withdrawn.

60. Pure Water Monterey. Accessed January 29, 2021. https://purewatermonterey.org.

61. Keller, A. A., Y. Su, and D. Jassby. "Direct Potable Reuse: Are We Ready? A Review of Technological, Economic, and Environmental Considerations." *ACS ES&T Engineering* 2, no. 3 (2022): 273–91.

62. Gerrity, "Potable Reuse Treatment Trains."

63. Voulvoulis, N. "Water Reuse from a Circular Economy Perspective and Potential Risks from an Unregulated Approach." *Current Opinion in Environmental Science and Health* 2 (2018): 32–45.

64. Marron, E. L., et al. "A Tale of Two Treatments: The Multiple Barrier Approach to Removing Chemical Contaminants during Potable Reuse." *Accounts of Chemical Research* 52 (2019): 615–22.

65. Du Pisani, "Direct Potable Reclamation."

66. Nappier, S. P., J. A. Soller, and S. E. Eftim. "Potable Water Reuse: What Are the Microbiological Risks?" *Current Environmental Health Risks* 5 (2018): 283–92.

67. Tortajada, "Drink More Recycled Wastewater."

68. Tennyson, P. A., M. Millan, and D. Metz. "Getting Past the 'Yuck Factor': Public Opinion Research Provides Guidance for Successful Potable Reuse Outreach." *Journal of American Water Works Association* 107 (2015): 58–62.

69. Furlong, C., et al. "Is the Global Public Willing to Drink Recycled Water? A Review for Researchers and Practitioners." *Utilities Policy* 56 (2019): 53–61.

70. WateReuse. *Framework for Direct Potable Reuse*. WateReuse, American Water Works Association, Water Environmental Federation, and National Water Research Institute, 2015. https://watereuse.org/wp-content/uploads/2015/09/14-20.pdf.

71. WHO, *Potable Reuse*.

72. National Research Council, *Understanding Water Reuse*.

73. US EPA. *Mainstreaming Potable Water Reuse in the United Stated: Strategies for Leveling the Playing Field*. April 2018.

74. Gerrity, "Potable Reuse Treatment Trains."

75. WHO, *Potable Reuse*.

CHAPTER 10

1. WHO/UNICEF Joint Monitoring Programme. "Progress on Drinking Water, Sanitation and Hygiene: 2017 Update and SDG Baselines." July 2017. https://data.unicef.org/resources/progress-drinking-water-sanitation-hygiene-2017-update-sdg-baselines/.

2. WHO/UNICEF Joint Monitoring Programme for Water Supply, Sanitation and Hygiene. *Progress on Household Drinking Water, Sanitation and Hygiene 2000–2020: Five Years into the SDGs*. World Health Organization (WHO) and the United Nations Children's Fund (UNICEF), 2021. https://washdata.org/sites/default/files/2021-07/jmp-2021-wash-households.pdf.

3. WHO. "Drinking-Water: Key Facts." Accessed February 6, 2020. https://www.who.int/en/news-room/fact-sheets/detail/drinking-water.

4. Hutton, G. "What Costs the World $260 Billion Each Year?" February 11, 2013. https://blogs.worldbank.org/water/what-costs-the-world-260-billion-each-year.

5. Workman, C., and H. Ureksoy. "Water Insecurity in a Syndemic Context: Understanding the Psycho-emotional Stress of Water Insecurity in Lesotho, Africa." *Social Science & Medicine* 179 (April 2017): 52–60.

6. UN. "General Assembly Adopts Resolution Recognizing Access to Clean Water, Sanitation as Human Right, by Recorded Vote of 122 in Favour, None

Against, 41 Abstentions." July 28, 2010. https://www.un.org/press/en/2010/ga10967.doc.htm.

7. Mintz, E., et al. "Not Just a Drop in the Bucket: Expanding Access to Point-of-Use Water Treatment Systems." *American Journal of Public Health* 91, no. 10 (2001): 1565–70.

8. Lantagne, D. S., R. Quick, and E. D. Mintz. *Household Water Treatment and Safe Storage Options in Developing Countries: A Review of Current Implementation Practices.* Wilson Center. https://www.wilsoncenter.org/sites/default/files/media/documents/publication/WaterStoriesHousehold.pdf.

9. WHO and UNICEF. *Progress on Household Drinking Water, Sanitation and Hygiene: 2000–2017.* 2019. https://www.unicef.org/media/55276/file/Progress%20on%20drinking%20water,%20sanitation%20and%20hygiene%202019%20.pdf.

10. WHO and UNICEF. *Progress on Household Drinking Water, Sanitation and Hygiene 2000–2020: Five Years into the SDGs.* 2021. https://www.who.int/publications/i/item/9789240030848.

11. WHO and UNICEF, *Drinking Water, Sanitation and Hygiene: 2000–2017.*

12. Dash, D. K. "Delhi's Tap Water Is Most Unsafe, Mumbai's Best." *Times of India*, November 17, 2019.

13. WHO, "Drinking-Water: Key Facts."

14. UNICEF. *Promotion of Household Water Treatment and Safe Storage.* UNICEF WASH Programmes, 2008.

15. WHO and UNICEF. *Safely Managed Drinking Water—Thematic Report on Drinking Water 2017.* 2017.

16. Lantagne, *Household Water Treatment.*

17. UNICEF, *Promotion of Household Water Treatment.*

18. WHO. *Combating Waterborne Diseases at the Household Level: The International Network to Promote Household Water Treatment and Safe Storage.* 2007.

19. WHO. *Evaluating Household Water Treatment Options: Health-Based Targets and Microbiological Specifications.* 2011.

20. WHO/UNICEF Joint Monitoring Programme for Water Supply, Sanitation and Hygiene. *Integrating Water Quality Testing into Household Surveys.* 2020.

21. WHO, *Combating Waterborne Diseases.*

22. Lantagne, *Household Water Treatment.*

23. Centre for Affordable Water and Sanitation Technology (CAWST). "Addressing the Water and Sanitation Challenge." Accessed February 6, 2020. https://www.cawst.org/why.

24. Huq, A., et al. "A Simple Filtration Method to Remove Plankton-Associated Vibrio Cholera in Raw Water Supplies in Developing Countries." *Applied and Environmental Microbiology* 62 (1996): 2508–12.

25. Colwell, R. R., et al. "Reduction of Cholera in Bangladeshi Villages by Simple Filtration." *Proceedings of the National Academy of Sciences* 100 (2003): 1051–55.

26. Huq, A., et al. "Simple Sari Cloth Filtration of Water Is Sustainable and Continues to Protect Villagers from Cholera in Matlab, Bangladesh." *mBio* 1 (2010): 1–5.

27. Huq, "Simple Filtration Method."

28. Huq, "Simple Filtration Method."

29. Colwell, "Reduction of Cholera."

30. CAWST. *Biosand Filter Construction Manual*. 2012.

31. Potters for Peace. "Ceramic Water Filter Program." Accessed February 6, 2020. https://www.pottersforpeace.org/ceramic-water-filter-project.

32. Hussam, A. "Contending with a Development Disaster: SONO Filters Remove Arsenic from Well Water in Bangladesh." *Innovations* 4, no. 3 (Summer 2009): 89–102.

33. Hussam, "SONO Filters."

34. Hussam, "SONO Filters."

35. Hussam, "SONO Filters."

36. Hussam, "SONO Filters."

37. Hussam, A., and A. K. M. Munir. "A Simple and Effective Arsenic Filter Based on Composite Iron Matrix: Development and Deployment Studies for Groundwater of Bangladesh." *Journal of Environmental Science and Health Part A* 42 (2007): 1869–78.

38. Neumann, A., et al. "Arsenic Removal with Composite Iron Matrix Filters in Bangladesh: A Field and Laboratory Study." *Environmental Science and Technology* 47 (2013): 4544–54.

39. WHO. *Results of Round I of the WHO International Scheme to Evaluate Household Water Treatment Technologies*. 2016.

40. WHO. *Results of Round II of the WHO International Scheme to Evaluate Household Water Treatment Technologies*. 2019.

41. LifeStraw. "About LifeStraw." Accessed February 24, 2020. https://www.lifestraw.com/pages/about-us.

42. LifeStraw. "How Our Products Work." Accessed February 24, 2020. https://www.lifestraw.com/pages/how-our-products-work.

43. Rosa, G., L. Miller, and T. Clasen. "Microbiological Effectiveness of Disinfecting Water by Boiling in Rural Guatemala." *American Journal of Tropical Medicine and Hygiene* 82 (2010): 473–77.

44. Clasen, T. F., et al. "Microbiological Effectiveness and Cost of Boiling to Disinfect Drinking Water in Rural Vietnam." *Environmental Science and Technology* 42 (2008): 4255–69.

45. Clasen, T., et al. "Microbiological Effectiveness and Cost of Disinfecting Water by Boiling in Semi-Urban India." *American Journal of Tropical Medicine and Hygiene* 79 (2008): 407–13.

46. Deep Springs International. "Home." Accessed February 6, 2020. http://www.deepspringsinternational.org/.

47. EAWAG. "SODIS Safe Drinking Water for All." Accessed February 6, 2020. http://www.sodis.ch/index_EN.

48. AJPU. "Guatemala Partnership." Accessed February 6, 2020. https://ncpguatemala.com/ajpu/.

49. Manob Shakti Unnoyon Kendro. "Welcome to Manob Shakti Unnoyon Kendro." Accessed February 6, 2020. http://www.msuk-bd.org/.

50. Potters for Peace. "Making Clean Water a Reality." Accessed February 6, 2020. https://www.pottersforpeace.org/.

51. Centre for Affordable Water and Sanitation Technology. "Vision and Mission." Accessed February 6, 2020. https://www.cawst.org/about/visionandmission/.

52. Rosa, "Rural Guatemala."

53. Clasen, "Rural Vietnam."

54. Clasen, "Semi-Urban India."

55. Makhuvele, R., and K. L. M. Moganedi. "Efficiency and Applicability of Low Cost Home-Based Water Treatment Strategies in a Rural Context." *Journal of Biological Sciences* 19 (2019): 339–46.

56. Makhuvele, "Efficiency and Applicability."

57. Rosa, "Rural Guatemala."

58. Firth, J., et al. "Point-of-Use Interventions to Decrease Contamination of Drinking Water: A Randomized, Controlled Pilot Study on Efficacy, Effectiveness, and Acceptability of Closed Containers, Moringa Oleifera, and in-Home Chlorination in Rural South India." *American Journal of Tropical Medicine and Hygiene* 82, no. 5 (2010a): 759–65.

59. Francis, M. R., et al. "Perception of Drinking Water Safety and Factors Influencing Acceptance and Sustainability of a Water Quality Intervention in Rural Southern India." *BMC Public Health* 15, no. 1 (2015): 731.

60. Water and Sanitation Program. *Improving Household Drinking Water Quality: Use of Biosand Filters in Cambodia.* 2010.

61. Workman, C. L. "Perceptions of Drinking Water Cleanliness and Health-Seeking Behaviours: A Qualitative Assessment of Household Water Safety in Lesotho, Africa." *Global Public Health* 14, no. 9 (2019): 1347–59.

62. Workman, "Perceptions of Drinking Water Cleanliness."

63. Workman, "Perceptions of Drinking Water Cleanliness."

64. Rojas, L. F. R., and A. Megerle. "Perception of Water Quality and Health Risks on Rural Area of Medellín." *American Journal of Rural Development* 1 (2013): 106–15.

65. Onjala, J., S. W. Ndiritu, and J. Stage. "Risk Perception, Choice of Drinking Water and Water Treatment: Evidence from Kenyan Towns." *Journal of Water, Sanitation and Hygiene for Development* 4 (2014): 268–80.

66. Francis, "Perception of Drinking Water Safety."
67. Masanyiwa, Z. S., I. J. E. Zilihona, and B. Kilobe. "Users' Perceptions on Drinking Water Quality and Household Water Treatment and Storage in Small Towns in Northwestern Tanzania." *Open Journal of Social Sciences* 7 (2019): 28–42.
68. CAWST HWTS Knowledge Base. "The Big Picture." Accessed May 27, 2022. https://www.hwts.info/the-big-picture.
69. Francis, "Perception of Drinking Water Safety."
70. Bitew, B. D., et al. "Barriers and Enabling Factors Associated with the Implementation of Household Disinfection: A Qualitative Study in Northwest Ethiopia." *American Journal of Tropical Medicine and Hygiene* 102 (2020): 458–67.
71. Francis, "Perception of Drinking Water Safety."
72. Rojas, "Perception of Water Quality."
73. Onjala, "Risk Perception."
74. Francis, "Perception of Drinking Water Safety."
75. Masanyiwa, "Users' Perceptions on Drinking Water Quality."
76. Bitew, "Barriers and Enabling Factors."
77. CAWST, "The Big Picture."
78. Centre for Affordable Water and Sanitation Technology. "There Is No Silver Bullet." Accessed February 6, 2020. https://www.hwts.info/experience/2ac6e1de/there-is-no-silver-bullet-technology.
79. Ndé-Tchoupé, A. I., et al. "White Teeth and Healthy Skeletons for All: The Path to Universal Fluoride-Free Drinking Water in Tanzania." *Water* 11 (2019): 131.
80. Ndé-Tchoupé, "White Teeth and Healthy Skeletons."
81. Gwenzi, W., et al. "Biochar-Based Water Treatment Systems as a Potential Low-Cost and Sustainable Technology for Clean Water Provision." *Journal of Environmental Management* 197 (July 2017): 732–49.
82. Ndé-Tchoupé, "White Teeth and Healthy Skeletons."
83. Department of Drinking Water & Sanitation, Ministry of Jalshakti. "Jal Jeevan Mission." Accessed January 10, 2022. https://jaljeevanmission.gov.in.
84. Jal Jeevan Mission. "Har Ghar Jal Dashboard." Accessed January 10, 2022. https://ejalshakti.gov.in/jjmreport/JJMIndia.aspx.
85. Mashal, M., and H. Kumar. "In India's Water-Stressed Villages, Modi Seeks a Tap for Every Home." *New York Times,* December 21, 2021. https://www.nytimes.com/2021/12/21/world/asia/india-water-modi.html.
86. Mashal, "India's Water-Stressed Villages."
87. Hutton, G. *Global Costs and Benefits of Drinking-Water Supply and Sanitation Interventions to Reach MDG Target and Universal Coverage.* World Health Organization, 2012.

CHAPTER 11

1. Smith, A. *An Inquiry into the Nature and Causes of the Wealth of Nations.* London: W. Strahan and T. Cadell, 1776.

2. Circle of Blue. "The Price of Water." Accessed March 13, 2020. https://www.circleofblue.org/waterpricing.

3. Food & Water Watch. "Tap Water vs. Bottled Water." Accessed March 13, 2020. https://www.foodandwaterwatch.org/about/live-healthy/tap-water-vs-bottled-water.

4. In the United States, municipal water is more tightly regulated than bottled water, which is regulated by the Food and Drug Administration (FDA). You can read more about bottled water regulations and how the quality of bottled water compares with municipal water at these sites: https://www.foodandwaterwatch.org/insight/take-back-tap-big-business-hustle-bottled-water, https://www.foodandwaterwatch.org/about/live-healthy/tap-water-vs-bottled-water.

5. Walton, B. "Price of Water 2017: Four Percent Increase in 30 Large U.S. Cities." May 18, 2017. https://www.circleofblue.org/2017/water-management/pricing/price-water-2017-four-percent-increase-30-large-u-s-cities/.

6. Layne, R. "Water Costs Are Rising Across the U.S.—Here's Why." August 27, 2019. https://www.cbsnews.com/news/water-bills-rising-cost-of-water-creating-big-utility-bills-for-americans.

7. Lakhani., N. "Revealed: Millions of Americans Can't Afford Water as Bills Rise 80% in a Decade." *The Guardian*, June 23, 2020. https://www.theguardian.com/us-news/2020/jun/23/millions-of-americans-cant-afford-water-bills-rise.

8. Water Education Foundation. "Water Rates." Accessed February 12, 2021. https://www.watereducation.org/topic-water-rates.

9. Walton, "Price of Water."

10. Patterson, L., and M. Doyle. *Ensuring Water Quality: Innovating on the Clean Water and Safe Drinking Water Acts for the 21st Century: A Report from the 2019 Aspen-Nicholas Water Forum.* 2019. https://www.aspeninstitute.org/wp-content/uploads/2019/11/WATER-Report-FINAL-002.pdf.

11. Legal Defense Fund. *Water/Color: A Study of Race and the Water Affordability Crisis in American Cities.* 2019. https://www.naacpldf.org/our-thinking/issue-report/economic-justice/water-color-a-study-of-race-and-the-water-affordability-crisis-in-americas-cities/.

12. Patterson, *Ensuring Water Quality.*

13. Patterson, L. A., and M. W. Doyle. *2020 Aspen-Nicholas Water Forum Water Affordability and Equity Briefing Document.* August 2020. https://nicholasinstitute.duke.edu/publications/2020-aspen-nicholas-water-forum-water-affordability-and-equity-briefing-document.

14. Legal Defense Fund, *Water/Color.*

15. Patterson, *2020 Aspen-Nicholas Water Forum.*
16. City of Santa Fe. "Water Rates." Accessed March 13, 2020. https://www.santafenm.gov/water_rates.
17. Gonzales, J. "Santa Fe Cuts Water Consumption by Imposing Tiered Pricing Model." Interview by Melissa Block, *All Things Considered*, NPR, May 13, 2015, audio. https://www.npr.org/2015/05/13/406505133/santa-fe-cuts-water-consumption-by-imposing-tiered-pricing-model.
18. Schwartz, N. D. "Water Pricing in Two Thirsty Cities: In One, Guzzlers Pay More, and Use Less." *New York Times,* May 6, 2015. https://www.nytimes.com/2015/05/07/business/energy-environment/water-pricing-in-two-thirsty-cities.html.
19. Legal Defense Fund, *Water/Color.*
20. City of Philadelphia. "Philadelphia Launches New, Income-Based, Tiered Assistance Program." June 20, 2017. https://www.phila.gov/press-releases/mayor/philadelphia-launches-new-income-based-tiered-assistance-program.
21. Food & Water Watch. "Historic Baltimore Water Justice Bill Becomes Law." Accessed February 12, 2021. https://www.foodandwaterwatch.org/news/historic-baltimore-water-justice-bill-becomes-law.
22. Baltimore City Department of Public Works. "Low-Income Water Bill Assistance Program." Accessed March 13, 2020. https://publicworks.baltimorecity.gov/low-income-water-bill-assistance-program.
23. Baltimore City Council. "Water Accountability and Equity Act." Accessed February 12, 2021. https://baltimore.legistar.com/LegislationDetail.aspx?ID = 3769175&GUID = 4A3F24AF-7CC7-442B-86C5-B01AF0A148F7.
24. Patterson, *2020 Aspen-Nicholas Water Forum.*
25. Garza, P. "California Gov. Gavin Newsom to Sign Bill to Create Safe and Affordable Drinking Water Fund." Environmental Defense Fund, July 24, 2019. https://www.edf.org/media/california-gov-gavin-newsom-sign-bill-create-safe-and-affordable-drinking-water-fund.
26. Klein., K. "Newsom Establishes Long-Term Safe and Affordable Drinking Water Fund." KVPR, July 25, 2019. https://www.kvpr.org/post/newsom-establishes-long-term-safe-and-affordable-drinking-water-fund.
27. Firestone, L., and S. De Anda. "Safe Drinking Water for All." *New York Times*, August 21, 2018. https://www.nytimes.com/2018/08/21/opinion/environment/safe-drinking-water-for-all.html.
28. Value of Water Campaign. "Imagine a Day without Water." Accessed March 13, 2020. http://imagineadaywithoutwater.org.
29. US EPA. "Water Conservation Tips for Residents." Accessed March 13, 2020. https://www3.epa.gov/region1/eco/drinkwater/water_conservation_residents.html.
30. New York City Department of Environmental Protection. "By Reducing Water Use During Rain Storms New Yorkers Can Help Protect the Health of

Local Waterways." March 5, 2018. https://www1.nyc.gov/html/dep/html/press_releases/18-014pr.shtml#.Yo0iLy-cbUY.

31. Levine, L. "A Wet 2018 Saw Sharp Rise in NYC Sewage Alerts: 1 in 3 Days." National Resources Defense Council, April 12, 2019. https://www.nrdc.org/experts/larry-levine/wet-2018-saw-sharp-rise-nyc-sewage-dumping-1-3-days.

32. US EPA. "Consumer Confidence Reports (CCR): Find Your Local CCR." Accessed March 13, 2020. https://ofmpub.epa.gov/apex/safewater/f?p=136:102.

33. Sloan, C. "The Unexpected Cause of Water Crises in American Cities." Talk Poverty, March 9, 2016. https://talkpoverty.org/2016/03/09/unexpected-cause-water-crises-american-cities/.

34. Kaffer, N. "About Half of Detroit Water Shutoffs Are Still Off." *Detroit Free Press*, January 13, 2020. https://www.freep.com/story/opinion/columnists/nancy-kaffer/2020/01/13/detroit-water-shutoffs/2834866001/.

35. Georgetown Climate Center. "Green Infrastructure Toolkit: Incentive-Based Tools." Accessed March 13, 2020. https://www.georgetownclimate.org/adaptation/toolkits/green-infrastructure-toolkit/incentive-based-tools.html.

36. vegetal i.D. "Green Roof Incentives." Accessed March 13, 2020. http://www.vegetalid.us/green-roof-technical-resources/green-roof-and-stormwater-management-incentives.html.

37. United Nations. "Clean Water and Sanitation: Why It Matters." Accessed January 2, 2021. https://www.un.org/sustainabledevelopment/wp-content/uploads/2016/08/6_Why-It-Matters-2020.pdf.

38. Kent, M. "Canadian Teen Tells UN 'Warrior Up' to Protect Water." CBC, March 22, 2018. https://www.cbc.ca/news/canada/autumn-peltier-un-water-activist-united-nations-1.4584871.

Additional Resources

SUGGESTED BOOKS

(Listed in order of publication date)

Rachel Carson. *Silent Spring.* Houghton Mifflin, 1962.
Philip Ball. *Life's Matrix: A Biography of Water.* University of California Press, 2001.
Charles Fishman. *The Big Thirst: The Secret Life and Turbulent Future of Water.* Free Press, 2011.
Alex Prud'homme. *The Ripple Effect: The Fate of Freshwater in the 21st Century.* Scribner, 2012.
Juliet Christian-Smith and Peter H. Gleick. *A Twenty-First Century US Water Policy.* Oxford University Press, 2012.
David Soll. *Empire of Water: An Environmental and Political History of the New York City Water Supply.* Cornell University Press, 2013.
Steven Mithen and Sue Mithen. *Thirst: Water and Power in the Ancient World.* Phoenix, 2013.
David Sedlak. *Water 4.0: The Past, Present, and Future of the World's Most Vital Resource.* Yale University Press, 2015.
James Salzman. *Drinking Water: A History.* Overlook Press, 2017.
Anna Clark. *The Poisoned City: Flint's Water and the American Urban Tragedy.* Metropolitan Books, 2018.

Seth M. Siegel. *Troubled Water: What's Wrong with What We Drink*. Thomas Dunne Books, 2019.

MEDIA

PBS. "H_2O: The Molecule That Made Us." Premiered April/May 2020. https://www.pbs.org/wgbh/molecule-that-made-us/.

LEARNING ACTIVITIES

Published Case Studies on Drinking Water

Below are examples of case study sites related to drinking water that extend ideas, concepts, and issues discussed in this book. The examples cut across disciplines, from the natural sciences to environmental governance and policy to social justice.

i) Case Studies in the Environment: https://online.ucpress.edu/cse   *Case Studies in the Environment* is a peer-reviewed journal for articles on environmental case studies and pedagogical articles on how to use case studies. The journal aims to inform faculty, students, researchers, educators, professionals, and policy makers on case studies and best practices in the environmental sciences and studies.

The cases listed below are examples that relate to drinking water. Search this site using the term *drinking water* to find additional cases.

Chitondo, Mandy, and Kelly Dombroski. "Returning Water Data to Communities in Ndola, Zambia: A Case Study in Decolonising Environmental Science." *Case Studies in the Environment* 3, no. 1 (December 2019): 1–8.

Hanlon, Jeffrey W. "Watershed Protection to Secure Ecosystem Services: The New York City Watershed Governance Arrangement." *Case Studies in the Environment* 1, no. 1 (December 2017): 1–6.

Lukacs, Heather A., Nik Sawe, and Nicola Ulibarri. "Risk, Uncertainty, and Institutional Failure in the 2014 West Virginia Chemical Spill." *Case Studies in the Environment* 1, no. 1 (December 2017): 1–7.

MacLeod, Clara, and Linda Estelí Méndez-Barrientos. "Groundwater Management in California's Central Valley: A Focus on Disadvantaged Communities." *Case Studies in the Environment* 3, no. 1 (December 2019): 1–13.

ii) National Center for Case Study Teaching in Science: https://www.nsta.org/case-studies   This site is a repository for peer-reviewed case studies for the natural sciences. While these cases are designed with a primary discipline in

mind, the cases on water draw from other disciplines. Below are a few examples of cases on this site related to drinking water.
Chester, Betty Jo, and Weslene T. Tallmadge. "What's in Your Water."
Homan, Michelle M. "The Water in Weberville."
Larrousse, Margaret M. "Setting Water on Fire: A Case Study in Hydrofracking."
Terry, Tracy J. "The Flint Water Crisis."

iii) Lessons in Conservation: https://ncep.amnh.org   This collection of peer-reviewed case studies is published by the Network of Conservation Educators and Practitioners.
Esbach, M., and M. Hedemark. "Payments for Ecosystem Services: An Introduction and Case Study on Lao PDR."
Weeks, B. C., and M. Esbach. "Valuing Ecosystem Services: A Qualitative Analysis of Drinking Water in the Solomon Islands."

Index

absorption, 121
acidity, measurement of, 180n38
activated carbon filtration, 56*fig.*, 57, 86, 121, 124*fig.*
Adenoviridae, 21*tab.*, 114*tab.*
adsorption, 121
advanced oxidation, 122, 123, 124*fig.*
affordable water access: governments and, 159; municipal water and, 157
Africa: cholera in, 22*fig.*; SODIS (solar disinfection) in, 148; Tanzania rainwater project, 151–52
agricultural sector: as contaminant source, 23, 29, 51*tab.*, 113*tab.*; fertilizers, 20, 26, 28–29, 67, 70, 78–81; impacts on water quality, 154; nitrate contamination and, 71; pollution from, 3; United Nations (UN), 80–81. *See also* nitrate contamination; pesticides
AJPU Association, 148
aldrin, 105
algal blooms, 2, 26–27, 67, 80, 128
aluminum sulfate, 55, 56*fig.*
ammonia: attraction, 11–13; boiling point, 13; hydrogen bonding, 13; liquid and gaseous states of, 15, 17; molecular structure of, 9–11, 12*fig.*; nitrate-based fertilizers and, 78–79, 99; physical properties of, 9*tab.*; polarity, 11–13, 17; in wastewater, 113*tab.*. *See also* liquid water
amoebic meningitis, 114*tab.*
androgenic hormones, 113*tab.*
anemia, 51*tab.*, 71, 80
antibiotics, 36, 112, 113*tab.*
antihypertensives, 113*tab.*
antiseptics, 113*tab.*
Appalachia, 2, 42
aquatic systems, 2, 16, 26, 80–81, 125, 128, 140, 148, 161–62. *See also* algal blooms
Argentina, arsenic in, 143–44
Arizona (USA), water projects in, 156–57
arsenic, 28, 42, 113*tab.*, 138, 143–46, 147*tab.*
Ascaris, 114*tab.*
Asia: cholera in, 22*fig.*; SODIS (solar disinfection) in, 148
Astroviridae, 21*tab.*, 114*tab.*
atmosphere, 7–8, 46*fig.*, 97, 99, 100
atrazine, 51*tab.*
attraction, 11–13
Australia, 84, 119*tab.*, 160*fig.*

209

bacterial diseases: cholera, 21*tab.*; dysentery, 21*tab.*, 114*tab.*; gastroenteritis, 21*tab.*, 114*tab.*; Guillain–Barré syndrome, 114*tab.*; respiratory illness, 21*tab.*, 114*tab.*; typhoid fever, 21*tab.*, 114*tab.*; as waterborne disease, 21*tab.*
bacterial pathogens: antibiotic-resistant bacteria, 112; *Campylobacter*, 114*tab.*; *Escherichia coli* (E. coli), 21*tab.*, 114*tab.*; *Legionella* spp., 21*tab.*, 114*tab.*; *Salmonella* Typhi, 21*tab.*, 114*tab.*; *Shigella*, 21*tab.*, 114*tab.*; *Vibrio cholerae*, 21*tab.*, 114*tab.*. *See also* pathogens
Baker City (Oregon, USA), 23
Baltimore (Maryland, USA), 158–59
Bangladesh: arsenic in, 143–44, 146; PFAS contamination, 84
basicity: calcium oxide and, 56; measurement of, 180n38; water treatment and, 57
basic needs: basic hygiene practices, 135; JMP and, 40*tab.*, 41*fig.*; lack of, 133*fig.*; safe drinking water and, xiv; tiered pricing and, 158
benzene, 51*tab.*
biochar, 151
biocides, 113*tab.*
bioswales, 107*tab.*
Bipartisan Infrastructure Bill of 2021, 78
boiling of water, 35, 137, 147*tab.*, 148, 149, 150
boiling points, 9, 10, 12–13
Bosch, Carl, 78, 79
bottled water: economics of, 156; regulation of, 156, 202n4
bromate, 113*tab.*
Buffalo (New York, USA), 159
built infrastructures, 102*tab.*

cadmium, 51*tab.*, 57, 113*tab.*
calcium, 55–56
calcium carbonate, 25–26, 57
calcium oxide, 56
Caliciviridae, 114*tab.*
California (USA): desalination projects, 127, 128; GWRS, 125; nitrate contamination, 2, 80–81; potable reuse and, 119*tab.*; Pure Water system, 127, 128, 131; Safe and Affordable Drinking Water Fund, 159; water projects in, 156–57
Campylobacter, 114*tab.*
Canada, 73, 148, 160*fig.*, 165–66
canals, 40*tab.*, 102*tab.*

cancer risks: chemical contamination and, 2, 51*tab.*, 112; EPA and, 93; nitrate contamination and, 80; PFOA contamination and, 2, 83; TCDD and, 27
Cape Fear River (North Carolina, USA), 85
carbon: activated carbon filtration, 56*fig.*, 57, 86, 121, 123, 124*fig.*; biogeochemical cycles and, 98; calcium carbonate, 25–26, 57; carbon dioxide, 6–7, 8*tab.*, 57, 99; chemical contaminants and, 51*tab.*; chlorofluorocarbons (CFCs), 81–82, 83*tab.*, 186n57; hydrogen bond with, 9; organic compounds, 99; in solar system, 6, 7; solubility and, 16–17, 25; TCCDs and, 27. *See also* methane
carbon dioxide, 6–7, 8*tab.*, 57, 99
careers in water sector, 163–64
Carson, Rachel, 48
Catskill-Delaware watershed protection program, 103–6
CAWST (Centre for Affordable Water and Sanitation Technology), 148, 151
CCL (Contaminant Candidate List), 65–66, 90
Centre for Affordable Water and Sanitation Technology (CAWST), 148, 151
ceramic pitcher filtration, 142–43, 147*tab.*
CFCs (chlorofluorocarbons), 81–82, 83*tab.*, 186n57
Charleston MCHM contamination (West Virginia, USA), 97
chemical bonding, 9–15, 78
chemical contaminants: in Charleston (West Virginia, USA), 97; education on, 162; in Long Island (New York, USA), 2; from mines, 2, 42, 51*tab.*, 172n4; natural infrastructure and, 102*tab.*; negative impacts of, 5, 164; radionuclides, 172n4; regulation of, 159; research and, 154; threats from, 2, 25–29; water contamination by, 81–87; water security objectives and, 102*tab.*. *See also* government regulatory action; PFAS contamination; PFOA contamination; PFOS contamination
chemistry of water: attraction, 11–13; dissolution in water, 164; electronegativity, 10–12; hydrogen bonding, 13, 14*fig.*; importance of, xiii–xiv; molecular formula of water, 9–13; paradox of water, 3, 4, 194; polarity, 12–13; precautionary principle and, 5; solid water, 14–16; vigilance and, xiv–xv, 29, 71, 155. *See also*

chemical contaminants; liquid water; molecular formula of water
Chemours, 85
Chile, 79, 143–44
China (PRC), People's Republic of, 106–7, 144
chloramine, 51*tab.*, 74
chlorate, 113*tab.*
chlorine: chloramine, 113*tab.*; chlorate, 113*tab.*; chlorination, 35, 37*fig.*, 52, 56*fig.*, 58, 72, 137–38, 146, 147*tab.*, 148, 150; chlorine dioxide, 51*tab.*; chlorite, 113*tab.*; chlorofluorocarbons (CFCs), 81–82, 83*tab.*, 186n57; as contaminant, 51*tab.*, 52–53; dioxin (TCDD), 27, 51*tab.*, 53; ferric chloride, 55, 56*fig.*; in filtration, 57; lead solubility and, 74–76; residual chlorine, 52, 58, 138; sodium chloride, 25–28. *See also* Flint water crisis (Michigan, USA); trihalomethanes (THMs)
chlorine dioxide, 51*tab.*
chlorite, 113*tab.*
chlorofluorocarbons (CFCs), 81–82, 83*tab.*, 186n57
cholera, 22*fig.*; global trends, 22*fig.*; in Haiti, 22–23; lead contamination and, 72; in Peru, 22; unfamiliarity with, 132; *Vibrio cholerae* pathogen, 21–22, 21*tab.*, 35, 114*tab.*; as waterborne disease, 4, 21–22, 21*tab.*, 114*tab.*
chromium, 51*tab.*, 57, 113*tab.*
Clark, Anna, 76
Clean Water Act (CWA): passage of, 50, 96, 154; regulatory issues, 159; sources of drinking water and, 54; wastewater discharge and, 58–60
Cleveland (Ohio, USA), 158
climate change: education on, 162; effects on water contamination, 66–67; potable reuse and, 117–18, 128; precautionary principle and, 5; precipitation patterns, 44, 161; stormwater runoff, 102*tab.*, 112, 161–62; stressing freshwater sources, 44; sustainability and, 153; wastewater treatment systems and, 161–62. *See also* greenhouse gases (GHGs)
climate resilience, potable reuse and, 128
cloth filtration, 138, 139, 139–40, 148
coagulation, 102*tab.*
coal mines, 2, 42, 51*tab.*
colloidal silver, 142, 143

combined-sewer systems, 161–62
compliance: economics of, 155, 157; public health and, 63–64
Consumer Confidence Reports, 162
Contaminant Candidate List (CCL), 65–66, 90
copper, 35, 57, 66, 73, 74, 77, 113*tab.*, 179n127
cost-benefit studies, 155
COVID-19 pandemic: effects of, 155; responses to, 155; water use during, xv–xvi
CREATE Initiative, 108
Croton Reservoir system, 103, 105
Cryptosporidium, 21*tab.*, 23, 58, 96, 114*tab.*, 122*fig.*, 138, 147*tab.*, 164
cultural perspectives: decentralization and, 5; water access and, xvi
Cultural-Resilience-Environment-Workforce (CREW), 108
CWA (Clean Water Act). *See* Clean Water Act (CWA)

dams, as built infrastructure, 102*tab.*
decentralization of water treatment: assessing HWTS impacts, 149–53; decentralized treatments, 134–36; global water usage and, 5, 132–34. *See also* point-of-use treatments
Deep Springs International, 148
de facto reuse (unplanned reuse), 112, 114, 115*fig.*
degreasers, 113*tab.*
delivery of water: chemistry and, xiv, xv, 3, 5, 19; infrastructure investments in, 4, 38–39; paradox of water and, 19; precautionary principle and, 5; regulatory issues for, 5; socioeconomic issues and, 25; threats to, 19–20, 44; vigilance and, 29. *See also* chemical contaminants; infrastructure issues; microbial contamination; piped water systems
desalination, 127, 128; RO and, 194n22
developed/developing distinction, 170n1
diamond-water paradox, 155–56
diarrheal diseases, 21, 51*tab.*, 137, 138, 139, 141, 143, 149. *See also specific diseases*
digestive issues, 51*tab.*, 71, 80
dilution effect, 114
dioxin (TCDD), 27, 51*tab.*, 53
direct potable reuse (DPR) systems, 116, 118*fig.*, 119*fig.*, 123*fig.*, 124, 126–27

diseases: dysentery, 21*tab.*, 114*tab*, 132; eye infections, 21*tab.*, 114*tab.*; infectious hepatitis, 21*tab.*; protozoal diseases, 21*tab.*; respiratory illness, 21*tab.*, 23–24, 48, 114*tab.*; typhoid fever, 4, 21*tab.*, 33, 35–36, 45, 72, 114*tab.*, 132; waterborne disease, 1–2, 4, 21*tab*. *See also* bacterial diseases; cholera; diarrheal diseases; gastroenteritis; protozoal diseases; viral diseases

disinfectant by-products, 113*tab.*; NPDWR and, 51*tab*.

disinfection: as contaminant source, 51*tab.*; ozone, 58; ultraviolet radiation, 58. *See also* chlorination

domestic waste, 154

drought, 156–57

Dupont Chemical Corporation, 82, 85

dysentery, 21*tab.*, 114*tab.*, 132

Earle, Sylvia, 165

Earth: ammonia on, 13; atmosphere of, 6–7, 8*tab.*; climate stability on, 30; life on, 16–17, 27; pure water on, 29; support of liquid water, 8–13, 15, 18, 28; water percentages on, 45, 46*fig*.

economics: of compliance, 155; cost-benefit studies, 155; of environmental regulations, 155; potable reuse and, 128; productivity increases, 4; of regulatory compliance, 165; value of safe drinking water, 159–64; water access and, xv, xvi, 1; of water improvement, 155; of water management, 155; 165; water rates, 163; of water treatment, 155. *See also* investment; investments; socioeconomic issues

ecosystem health: Catskill-Delaware watershed protection program, 103–6; green infrastructures and, 106–9; precautionary principle and, 155; protecting watersheds, 98–103; protections for, 96–98; regulatory infrastructure and, 159; safe drinking water and, 165; watershed protection programs, 97–98

education: investments and, 155; safe drinking water and, 165; on source to tap process, 162–63

electronegativity, 10–12

electrons, 9–13

Elk River MCHM contamination (Charleston, West Virginia), 97

emissions from forest fires, 3, 20, 67, 100

engineering infrastructure, 154

environmental issues; economics of, 155; potable reuse and, 128; precautionary principle and, 155; public activism and, 159. *See also* ecosystem health; watersheds

EPA (Environmental Protection Agency): education on, 162; on GenX contaminant, 85; PFAS contamination and, 83–87, 89; on PFAS regulation, 86–87; on PFOA/PFOS regulation, 86; on public water systems, 180n39; role of, 4; TSCA and, 92–94. *See also* Safe Drinking Water Act (SDWA)

epoxy resins, 113*tab*.

equitable access, xvi, 157, 159

erosion control, 102*tab*.

Escherichia coli (E. coli), 21*tab.*, 51*tab.*, 114*tab*.

estradiol, 105, 113*tab*.

estrogenic hormones, 113*tab*.

estrone, 113*tab*.

European Chemicals Agency (ECHA), 94

European Union (EU): 160*fig.*; cholera in, 22*fig.*; ECHA and, 94; PFAS contamination, 84; REACH law, 92

eye infections/irritations, 21*tab.*, 48, 51*tab.*, 114*tab*.

FDA (Food and Drug Administration) bottled water regulation, 202n4

fecal contamination, 21, 51*tab.*, 149

Federal Water Pollution Control Act, 50. *See also* Clean Water Act (CWA)

ferric chloride, 55, 56*fig*.

fertilizers, 20, 26, 28–29, 67, 70, 78–81. *See also* nitrate contamination

filtration: in ancient periods, 31, 35; biochar, 151–52; biosand filtration, 140–42, 142, 147*tab.*; Catskill-Delaware watershed and, 103–5; ceramic pitcher filtration, 142–43, 147*tab.*; cloth filtration, 139–40, 148; education on, 162; EPA and, 86, 105; FADs, 105; green infrastructures and, 106, 106–7; Hazen and, 36; high/low pressure, 121–22; iron filtration, 151–52; membrane filtration, 102*tab.*, 121–23, 123; multibarrier approach and, 122–24, 126, 128, 129, 130; natural filtration process, 94, 106; pore sizes and, 122; potable reuse and, 121; sand filtration, 31, 35; slow sand filtration, 151–52; SONO arsenic filtration, 143–46, 147*tab.*; in source to tap path, 56–57; surface water treatment and, 104;

INDEX 213

urban systems of, 72; water hardness, 180n37. *See also* nanofiltration; reverse osmosis (RO)
filtration avoidance determinations (FADs), 105
firefighting foams, 82, 83, 84–85, 113*tab.*
Flint water crisis (Michigan, USA), xv, 2, 25, 66, 67, 71, 73–78, 130, 159
floodwaters, 66–67, 102*tab.*
fluorine/fluorinated compounds, 56, 81–83, 83*tab.*, 85, 113*tab.*. *See also* PFAS contamination; PFOA contamination; PFOS contamination
Food and Drug Administration (FDA) bottled water regulation, 202n4
food security, 165
forests: fires, 3, 20, 67, 100, 102*tab.*.
forward-thinking, 5. *See also* precautionary principle
Fountain (Colorado, USA), 71
freezing, 7, 8, 14–16
freshwater sources: osmosis and, 194n22; potable reuse and, 128
Fuller, Thomas, 165

gastroenteritis, 21*tab.*, 114*tab.*. *See also* *Cryptosporidium*; *Giardia*
gastrointestinal illnesses, 23–24, 51*tab.*. *See also specific diseases*
gender equity, safe drinking water and, 165
GenX contamination, 71, 83*tab.*, 85, 97
GHGs (greenhouse gases), 6–7, 91, 127. *See also* carbon dioxide; climate change; methane
Giardia, 21*tab.*, 51*tab.*, 58, 114*tab.*, 122*fig.*, 147*tab.*
Glennon, Robert, 158
global economics, 157–58
Global North: term usage, 170n1; water access, xv, 1; water usage, 160–61
Global South: safe drinking water access, 132–34; term usage, 170n1; water access, xv. *See also* decentralization of water treatment
Global Water Project, 108
global water usage, 160*fig.*
government regulatory action, 154. *See also* legislative issues
greenhouse gases (GHGs), 6–7
green infrastructures, 106–9
groundwater contamination: in Appalachia, 2; in Long Island, 2; water infiltration and, 102*tab.*

Ground Water Replenishment System (GWRS), 125
guano fertilizer, 79
Guatemala, 142, 148, 149
Guillain-Barré syndrome, 114*tab.*

Haber, Fritz, 78, 79, 185n44
Haber-Bosch process, 78–80
haloacetic acids, 51*tab.*, 113*tab.*
hardness of water, 55–56. *See also* pH measurements
Hazen, Alan, 36
heat capacity, 15
heavy metals. *See* metal toxicity
helminths, 114*tab.*
hepatitis, infectious, 21*tab.*, 114*tab.*
Hepeviridae, 21*tab.*, 114*tab.*
home water treatment and safe storage (HWTS) designs. *See* HWTS (home water treatment and safe storage) methods
Hoosick Falls PFAS contamination (New York, USA), 2, 25, 67, 71, 130, 159
household products, 48, 89, 113*tab.*
human/animal waste, 51*tab.*, 79, 113*tab.*
human rights and safe drinking water, 165
Hussam, Abul, 144, 148
HWTS (home water treatment and safe storage) methods: assessing impacts on community health, 149–53; defined, 136; economics of, 147*tab.*, 150, 151, 152; examples of, 137–48; external agencies and, 148; key factors of, 136; WHO on, 136, 146, 150. *See also* point-of-use treatments
hydrogen: hydrogen peroxide, 122; lead and, 74; paradox of water and, xiv; properties of water and, xiv, 3, 18, 20, 27, 29; recycling of elements and, 98
hydrogen bonding, 9–15, 78
hydrologic cycle, 15, 79, 110, 111*fig.*, 116, 128, 129–30

ice. *See* solid water
India: arsenic in, 143–44; HWTS methods and, 149; Jal Jeevan Mission, 152–53; nitrate contamination, 80; PFAS contamination, 84
indigenous communities, xv–xvi, 2, 5, 31, 42, 49, 98, 109, 148, 165–66
indirect potable reuse (IPR) systems, 116, 117*fig.*, 119*fig.*, 123*fig.*, 124, 127
Indonesia, PFAS contamination, 84

INDEX

industrialization: contamination and, 165; impacts on water quality, 154; pollution from, 3

industrial sector: as contaminant source, 51*tab.*, 71, 113*tab*. *See also* Hoosick Falls PFAS contamination (New York, USA); PFAS contamination

inequality of access, 155, 157

infectious diseases, 154. *See also* diseases

infectious hepatitis, 21*tab.*, 114*tab.*

infiltration, 99*fig.*, 102*tab.*, 106, 107*tab.*

infrastructure issues, 102*tab.*; built infrastructures, 102*tab.*; economics of, 156–57; education on, 162; engineering infrastructure, 154; investments and, 154; legislation and, 4–5; maintenance, 156; natural infrastructures, 102*tab.*, 164; promoting green infrastructure, 163; regulatory infrastructure, 154; scientific infrastructure, 154; socioeconomic issues and, 134. *See also* built infrastructures; engineering infrastructure; natural infrastructure; regulatory infrastructure; scientific infrastructure

inorganic chemicals: as contaminant source, 113*tab.*; NPDWR and, 51*tab.*; in wastewater, 113*tab.*. *See also* ammonia; fluoride; nitrate contamination

An Inquiry into the Nature and Causes of the Wealth of Nations (Smith), 155–56

investments: cost-benefit studies, 155; in early 20th century, 4; in infrastructures, 5, 154; need for, 3, 155

ion exchange process, 57, 74, 81, 86, 180n37

ionic compounds, 27–29, 55, 56*fig.*, 74, 80

Iowa (USA), 80–81

iron ions, 55, 56*fig.*, 57, 72

iron matrix filter, 144–45, 151

Jal Jeevan Mission, 152–53

Japan, 84, 160*fig.*

Joint Monitoring Programme for Water Supply, Sanitation and Hygiene (JMP), 40–42, 132, 135*fig.*

Kansas (USA), 116, 126

kidney toxicity, 51*tab.*

Latin America: cholera in, 22–23; SODIS (solar disinfection) in, 148. *See also specific countries*

leaching: of arsenic, 144; chloride levels and, 74, 76; defined, 57; from landfills, 65; of lead, xv, 56*fig.*, 57, 73–74, 77; of mercury, 42; pH levels and, 57; prevention of, 74; of trichloroethylene, 2

Lead and Copper Rule, 66, 73, 74, 77, 179n27

lead contamination: cases of, 155; chloride levels and, 74, 76; in early plumbing, 28, 58; effects of, 71–72, 73; leaching from pipes, xv, 56*fig.*, 57–58, 73, 74, 77; Lead and Copper Rule, 77, 179n27; NPDWR and, 51*tab.*; redlining and, 76–77; in Roman Empire, 72; SDWA and, 73; skepticism about, 72–73; US events of, 25; in Washington, DC, 74; in wastewater, 113*tab.*. *See also* Flint water crisis (Michigan, USA); piped water systems

Lead Industries Association (LIA), 72

Legionella spp., 21*tab.*, 48, 51*tab.*, 114tab.

legislative issues: Bipartisan Infrastructure Bill of 2021, 78; in California, 159; Clean Water Act (CWA), 154; infrastructure and, 4–5; in regulating water quality, 4; Restriction of Chemicals (REACH) law, 92–94; Safe Drinking Water Act (SDWA), 50, 154; Source Water Protection Program, 54, 59*fig.*, 96–97; Toxic Substances Control Act (TSCA), 92–93, 159. *See also* government regulatory action

levees, 102*tab.*

LIA (Lead Industries Association), 72

LifeStraw filter system, 146, 148

liming, 56

liquid water: about, 6–7; boiling point of, 9*tab.*, 13; earth's support of, 7, 8–13; molecular structure of, 9–13, 14*fig.*; properties of water, 9*tab.*, 14–16; as universal solvent, 16–18

Madrid Statement, 84

magnesium, 55–56

Manob Shakti Unnoyon Kendro (NGO), 148

marginalization: safe drinking water and, xv; water contamination and, 2

Mars, 6–7, 8*tab.*, 45

maximum contaminant level goals (MCLGs), 53–54, 64

maximum contaminant levels (MCLs), 53, 63–64, 65, 80, 84, 86, 88, 89

Mazariegos, Fernando, 142

MCHM contamination, 97

MCLGs (maximum contaminant level goals), 53–54, 64

MCLs (maximum contaminant levels), 53, 63-64, 65, 80, 84, 86, 88, 89
MDGs (Millennium Development Goals), 39-40
melting points, 9, 10, 14-15, 71-72
membrane filtration, 102*tab.*, 121-23
mental development issues, 51*tab.*, 52, 71-72, 91, 112
mercury (element), 42, 51*tab.*, 100, 113*tab.*
Mercury (planet), 6
metal toxicity, 2, 42, 57, 113*tab.*. *See also* arsenic; cadmium; chromium; copper; lead contamination; mercury; nickel; silver
methane, 6-11, 12*fig.*, 13, 15, 17
methylcyclohexane methanol (MCHM), 97
Michigan Civil Rights Commission report, 76
microbial biodegradation, 102*tab.*
microbial contamination: filtration and, 121-23; infectious diseases and, 154; NPDWR and, 51*tab.*; threats from, 20-25
microfiltration (MF), 121, 124*fig.*
microorganisms, role in chemical breakdown, 20-21
Millennium Development Goals (MDGs), 39-40
Millennium Ecosystem Assessment Board, 98
Milwaukee (Minnesota, USA), 23
mines/mining, 2, 42, 51*tab.*, 172n4
Minnesota (USA), 23, 80-81, 108
Mississippi (USA), 81
moisture retention, 102*tab.*
molecular formula of water, 9-13; about, 4; essential for life and, 4; properties of water and, 4
mortality rates, decline of, 4, 36, 52
municipal water: economics of, 156-57; education on, 162-63; regulation of, 202n4
Munir, Abul, 144, 148

Naegleria fowleri, 114*tab.*
Namibia (Africa), 119*tab.*, 126-27, 129, 131
nanofiltration (NF), 57, 86, 121-22
nanomaterials, 113*tab.*. *See also* silver
national standards, 162
Native Americans, xv-xvi, 2, 5, 42
native grasslands, 102*tab.*
natural infrastructure, 164, 101-3
Neosho River (Chanute, Kansas), 116
Nepal, 84
neurological issues, 51*tab.*, 52, 71-72
Newark water contamination (New Jersey, USA), 25, 67, 71, 73, 74, 77, 78, 159

NEWater system, 125, 131
New Mexico (USA), 158
New York (USA): Buffalo, 159; Catskill-Delaware watershed protection program, 103-6; Hoosick Falls PFAS contamination, 25, 159
New York City Watershed Agreement, 104-6
NGOs (nongovernmental organizations), 148
nickel, 113*tab.*
nitrate contamination: in advanced water treatment, 121; algal blooms and, 80; cases of, 155; climate change and, 67; in conventional water treatment, 120; fertilizers and, 70, 71, 78, 78-81, 79-81; filtration and, 57; NPDWR and, 51*tab.*; regulation of, 87, 97, 155; in runoff, 26, 28, 106, 114; in San Joaquin Valley, 2, 71; sources of, 26, 28, 51*tab.*; in wastewater, 113*tab.*; watersheds and, 99, 101
nitrite, 80, 113*tab.*, 121
nitrogen: in air, 20, 78; in ammonia, 9; in atmosphere, 8; atmospheric nitrogen, 78-79, 99; electronegativity of, 101; hydrogen bonding, 9; nitrate-based fertilizers, 79-80; nitrogen cycles, 98-100; in soil, 79. *See also* nitrate contamination
N-nitrosodimethylamine, 113*tab.*
nongovernmental organizations (NGOs), 148
nonpolar, 11, 12*fig.*, 13, 17, 18*fig.*, 26-27
nonsteroidal anti-inflammatories (NSAIDs), 113*tab.*
North Carolina (USA) 71, 83*tab.*, 85, 97

Oregon (USA), 23
organic chemicals, 51*tab.*
organic compounds, 99
organic matter in soils, 102*tab.*
osmosis, 194n22
Östersund (Sweden), 23
oxygen: in atmosphere, 8*tab.*; biogeochemical cycles and, 98; chlorine dioxide, 51*tab.*; electronegativity of, 10-11, 12*fig.*, 13*fig.*; hydrogen bond with, 9-10, 14*fig.*; hydrogen peroxide, 122; nitrate contamination and, 80; paradox of water and, xiv; properties of water and, xiv, 3, 18, 20, 26, 27, 29; wastewater treatment and, 120. *See also* carbon dioxide
ozone for disinfection, 58

paradox of value, 155-56
paradox of water, xiv, 3, 4, 18, 19-20, 29, 164-65

Parkersburg, (West Virginia, USA), 71
pathogenic contamination: impacts of, 23–25; natural infrastructure and, 102*tab.*; water security objectives and, 102*tab.*. *See also* Cryptosporidium
pathogens: advanced water treatment and, 121–25; bacteria, 21*tab.*, 147*tab.*; biosand filtration and, 140–42, 147*tab.*; boiling and, 137, 147*tab.*; CAWST and, 151–52; ceramic pitcher filtration and, 142–43, 147*tab.*; chlorination and, 137–38, 147*tab.*; cloth filtration and, 139–40; as contaminants, 51*tab.*, 102*tab.*; deactivating, 29; disinfection and, 56*fig.*, 58; germ theory and, 35; HWTS methods and, 146–48, 146–50; from livestock farms, 97; piped water systems and, 72; precautionary principle and, 96; protozoa, 21*tab.*, 147*tab.*; recontamination by, 135–36, 137, 138, 139, 140, 142, 143, 146; residual pathogens, 52, 56*fig.*, 58; SDWA and, 60; SODIS and, 138–39, 147*tab.*; SONO arsenic filtration and, 143–46, 147*tab.*; stormwater runoff and, 67; viruses, 21*tab.*, 147*tab.*; in wastewater, 112, 114*tab.*; water and sewage treatments, 23, 35–38, 52, 72; waterborne, 21–25. *See also* bacterial pathogens; cholera; *Cryptosporidium*; *Giardia*
Peltier, Autumn, 165–66
perchlorate, 88–89, 90, 91, 159
perfluoroalkyl substances. *See* PFAS contamination
Perfluorooctanesulfonic acid (PFOS). *See* PFOS contamination
perfluorooctanoic acid (PFOA). *See* PFAS contamination
personal care products, 29, 48, 65, 90, 113*tab.*
Peru, 22
pesticides, 48, 90, 113*tab.*; ecosystem health and, 48, 65; precautionary principle and, 90; in runoff, 20, 67, 97, 101, 114; in San Joaquin Valley, 2; testing of, 105; treatment for, 138; in wastewater, 113*tab.*
PFAS contamination: cases of, 155; EPA and, 83–87, 86–87, 89; in Hoosick Falls (New York, USA), 25; impacts of, 164; potable reuse and, 128; precautionary principle and, 90, 91; regulation of, 93–94, 155, 159; testing for, 105; treatment for, 123; in USA, 71; in wastewater, 113*tab.*. *See also* Hoosick Falls PFAS contamination (New York, USA)

PFOA contamination, 2, 25, 67, 82–86, 113*tab.*
PFOS contamination, 82–86, 113*tab.*
pharmaceuticals, 29, 48, 65, 90, 113*tab.*
Philadelphia (Pennsylvania, USA), 158–59
pH measurements: adjusting of, 56, 57; anticorrosion treatment and, 57, 73–74; defined, 180n38; nitrate contamination and, 80
phospholipid molecules, 17, 18*fig.*, 70
phthalates, 113*tab.*
piped water systems: anticorrosion treatment, 56*fig.*, 57–58; CAWST and, 151; cholera epidemic and, 34; during freezing temperatures, 16; heavy metals in, 113*tab.*; HWTS methods and, 150; in India, 134, 152–53; indigenous communities and, 42; lack of, xvi, 2, 134; lead contamination, xv, 25, 28, 57–58, 59, 71–78, 179n27; population growth and, 32; regulation of, 73, 74, 77–78, 180n39; Roman use of, 72; socioeconomic issues and, 134, 157, 159; source to tap path, 3, 54–60, 56*fig.*; water treatment and, 35, 52, 135–36. *See also* Flint water crisis (Michigan, USA); lead contamination; public water systems
plant uptake, as natural infrastructure, 102*tab.*
plasticizers, 113*tab.*
plumbing and pipes. *See* piped water systems
Plunkett, Roy, 81–82
point-of-use treatments: biosand filtration, 140–42; boiling, 137; ceramic pitcher filtration, 142–43; chlorination, 137–38; cloth filters, 139–40; external agencies and, 148; solar disinfection (SODIS), 138–39; SONO arsenic filtration, 143–46, 148. *See also* HWTS (home water treatment and safe storage) methods
polarity, 10, 11–13, 14, 15, 16–18, 26–27. *See also* hydrogen bonding
political issues: affordable water access, 159, 165; contamination and, 165; environmental regulations and, 155; equitable access, 159; voting, 164; water access and, xvi
pollution prevention, 102*tab.*
polychlorinated biphenyls, 113*tab.*
polyfluoroalkyl substances. *See* PFAS contamination
polymers, 82, 113*tab.*

polytetrafluoroethylene (PTFE) (Teflon), 82, 83*tab.*
POPs (persistent organic pollutants), 83. *See also* PFOA contamination; PFOS contamination
postcontamination treatments, 154
Postel, Sandra, 108
potable reuse: advanced water treatment for, 121–25; advantages of, 128; ancient water recycling, 111; challenges of, 129–30; conventional wastewater treatment for, 120; defined, 111; examples of, 125–28; hydrologic cycle and, 110–11; planned potable reuse, 111, 116–18, 119*tab.*; precautionary principle and, 111–12; public engagement and, 131; state of systems of, 130–31; from waste to potable reuse, 119–20; wastewater discharge and, 112, 114
Potters for Peace (NGO), 148
poverty reduction, 165
precautionary principle: application of, 92; chemical regulation and, 92–94; defined, 90–91; for environmental regulations, 155; importance of, 5; perchlorate and, 88–89; scientific data and, 91–92; SDWA and, 88–90; water management and, 94–96. *See also* ecosystem health; potable reuse systems
precipitation: hydrologic cycle, 15, 79, 110, 111*fig.*, 116, 128, 129–30; patterns of, 31, 44, 66, 128, 161; water quality and, 3; watersheds and, 98, 99*fig.*
protons, 9–10
protozoa, 21*tab.*; *Cryptosporidium*, 21*tab.*, 23, 58, 96, 114*tab.*, 122*fig.*, 138, 147*tab.*, 164; *Giardia*, 21*tab.*, 51*tab.*, 58, 114*tab.*, 122*fig.*, 147*tab.*; *Naegleria fowleri*, 114*tab.*
protozoal diseases, 21*tab.*, 114*tab.*
PTFE (polytetrafluoroethylene) (Teflon), 82, 83*tab.*
public health: COVID-19 pandemic and, xv–xvi, 155; overflow events and, 161–62; precautionary principle and, 155; risks to, xvi; safe drinking water and, xiv, xv, 165; SDWA and, 69; unsafe water consumption and, 1–2, 4; vigilance and, 159; water access and, 1. *See also* infectious diseases
public vigilance: about, 5; chemistry and, 3; education, 162–63; green infrastructure promotion, 163; role of public, 159–64; voting, 164; water conservation, 160–92; water contamination and, 71; water rates understanding, 163; water-related careers, 163–64
public water systems: EPA on, 180n39. *See also* Flint water crisis (Michigan, USA); piped water systems; water access
pure water, 28, 29, 73, 96, 121, 123, 125, 130, 180n38
Pure Water system, 127, 128, 131

quality control: chemistry and, 3. *See also* water quality

race issues: inequality of access, 165; redlining, 76
radionuclides, 172n4; NPDWR and, 51*tab.*
radium, 51*tab.*, 172n4
rainfall, 102*tab.* *See also* climate change; runoff
rain gardens, 107*tab.*
REACH (Restriction of Chemicals) law, 92–94
regulations: of bottled water, 202n4; framing of, 164; LIA and, 72; precautionary principle and, 155; on use of lead, 72. *See also* environmental regulations
regulatory infrastructure: about, 4–5; contamination and, 155; economics of, 157; investments in, 154; pace of regulatory processes, 164–65
relative mass (amu), 9
reproductive issues, 27, 51*tab.*, 71–72, 80, 93, 112
reservoirs: as built infrastructure, 102*tab.*; IPR and, 123*fig.*, 124; NEWater and, 125; New York City and, 103–4, 105; as outflow point, 98; water quality and, 3
residual chlorine, 52, 58, 138
respiratory illness, 21*tab.*, 23–24, 48, 114*tab.*
Restriction of Chemicals (REACH) law, 92–94
reverse osmosis (RO), 57, 86, 102*tab.*, 121–22, 124*fig.*, 194n22
Right to Water Coalition (Baltimore, USA), 158–59
Roman Empire contamination event, 70, 72
root systems, 102*tab.*
runoff: aquatic systems and, 26–27; nitrate contaminant in, 26, 28, 106, 114; overflow events and, 161–62; reduction of, 102*tab.*

Safe and Affordable Drinking Water Fund (California, USA), 159
safe drinking water: about, 45–46; access to, xv–xvi, 3, 164; basic needs and, xiv; beneficiaries of, 155; chemistry and, xiv, xv, 3; climate change and, 66–67; compliance costs, 63–64; economics of, 157; expectation of, 154; in Global North, xv, 1; inequality of access, 62–63; infrastructure investments, 67–69; NPDWR and, 50–54; oversight and, 155; public health and, xiv; quotes on, 165–66; regulating for, 47–50; regulatory challenges, 65–66; SDWA and, 60–62; source to tap path, 54–60; synthetic chemical contaminants, 64–65; value of, 155; vigilance and, 71, 155; WHO definition of, 1. *See also* water contamination
Safe Drinking Water Act (SDWA), 154; amendments to, 73, 96; CCL and, 65–66, 90; distribution and, 59*fig.*; Lead and Copper Rule, 66, 73, 74, 77, 179n27; MCLs and, 89–90; municipal water reports, 162; NPDWR and, 50–54; passage of, 50, 96, 154; pathogens and, 60–62; perchlorate and, 88; PFAS contaminants and, 86; regulatory issues, 159; Source Water Protection Program, 54, 59*fig.*; successes of, 60–62, 69; UCMR and, 66, 90; vigilance and, 159; water quality and, 155; water treatment, 59*fig.*
Salmonella Typhi, 21*tab.*, 114*tab.*
salt (sodium chloride), 25–28
saltpeter (potassium nitrate), 79
salts, dissolved, 7, 28, 83*tab.*, 85
saltwater, 28, 125, 194n22
San Joaquin Valley (California, USA), 2, 81
scarcity of water, 116–17
scientific infrastructure, 154
Scotchgard stain repellant, 82
SDWA (Safe Drinking Water Act). *See* Safe Drinking Water Act (SDWA)
sediment removal, 102*tab.*
sewage: as contaminant source, 23; overflow events, 161–62. *See also* wastewater treatment systems
Shigella, 21*tab.*, 114*tab.*
Silent Spring (Carson), 48
silting of waterways, water security objectives and, 102*tab.*
silver, 35, 71, 113*tab.*, 142, 143
Singapore, 106, 125, 131

Smith, Adam, 155–56
socioeconomic issues: agricultural sector, 80–81; decentralization and, 5; flooding, 108; HWTS methods and, 147*tab.*, 148; inequality of access, 165; infrastructure and, 95, 134; investments and, 155; regulation violations, 62–63, 81; safe drinking water and, xv, 165; tiered pricing, 158–59; value of safe drinking water, 159–64; water access and, xv, xvi, 1; water contamination and, 2, 5, 25
SODIS (solar disinfection), 138–39, 147*tab.*, 148
sodium chloride, 25–28
sodium hypochlorite, 137, 147*tab.*
soil composition, 100–101; as natural infrastructure, 102*tab.*
Soil Science Society of America, 100–101
solar disinfection (SODIS), 138–39, 147*tab.*, 148
solid water (ice), 14–16
solubility: acidity/basicity of solutions, 180n38; ionic compounds and, 27–29; of lead, xv, 56*fig.*, 57–58, 73–74, 76, 77; levels of, 25–29; pH and, 180n38; TCDD (dioxin) and, 27, 51*tab.*, 53; of toxins, 2
solvents, 16–18; universal solvent, 25–29
SONO arsenic filter, 143–46, 147*tab.*, 148
sources of drinking water, 56*fig.*; about, 3, 4, 154; education on, 162; origination point of, 3; pollution of, 4; precautionary principle and, 155; source to tap path, 4, 54–60
Source Water Protection Program, 54, 59*fig.*, 96–97
South Africa: DPR systems in, 127; HWTS methods and, 149; potable reuse and, 119*tab.*
Sponge City initiative, 106–7
statins, 113*tab.*
steroidal hormones, 105, 113*tab.*
Stockholm Convention, 83, 84
storage: as built infrastructure, 102*tab.*; chlorination and, 138
styrene, 51*tab.*
sugar (sucrose), 16, 25–26, 27
Surface Water Treatment Rule, 104
sustainability, xvi
Sustainable Development Goals (SDGs), xiv, 39–42
Sweden, 23
Swiss Federal Institute of Aquatic Science and Technology, 148

synthetic industrial chemicals, 48, 64–65, 70–71, 79–81, 113*tab.*
Szent-Györgyi, Albert, 165

Taenia, 114*tab.*
Tanzania (Africa), 151–52
TCDD (dioxin), 27, 51*tab.*, 53
TEFLON (polytetrafluoroethylene) (PTFE), 82, 83*tab.*
testosterone, 113*tab.*
Texas (USA): potable reuse and, 119*tab.*, 127; Trinity River, 114–15; water projects in, 156–57
Thailand, PFAS contamination, 84
THMs (trihalomethanes), 51*tab.*, 75, 113*tab.*, 138
threats from water: about, 19–20; chemical contaminants, 25–29; microbial contamination, 20–25
3D representations, 9–10
3M Corporation, 82–83
Tiered Assistance Program (TAP), 158–59
tiered pricing, in Santa Fe (New Mexico, USA), 158
Titan (moon), 8, 17
titanium oxide, 113*tab.*
Toxic Substances Control Act (TSCA), 92–93, 159
transportation of water, 15, 47, 116, 154. *See also* delivery of water
triclocarban, 113*tab.*
triclosan, 113*tab.*
trigonal pyramidal structure, 10, 11
trihalomethanes (THMs), 51*tab.*, 75, 113*tab.*, 138
TSCA, regulatory issues, 159
TSCA (Toxic Substances Control Act), 92–93, 159
turbidity, 31, 55, 106, 139, 140, 141, 142, 145, 147*tab.*, 149
typhoid fever, 4, 21*tab.*, 33, 35–36, 45, 72, 114*tab.*, 132

ultrafiltration (UF), 121, 124*fig.*
ultraviolet radiation, 58
UNEP (United Nations Environment Programme), 83
UNICEF (United Nations Children's Fund), 19, 38, 40–42, 132, 135*fig.*
United Nations (UN): Haitian cholera outbreak and, 22–23; Millennium Development Goals (MDGs), 39, 39–40; Peltier's address to, 165–66; recycled potable water, 112; Sustainable Development Goals (SDGs), xiv, 39–42; UN Children's Fund (UNICEF), 19, 38; UN Environment Programme (UNEP), 83; UN Food and Agriculture Organization report, 80–81; on water access, xiv, 1, 133
United States: arsenic in, 143–44; chemical regulation and, 92–94; *Cryptosporidium* contamination in, 23; water contamination in, 2–3, 71; water pricing structures in, 158; water rates in, 157–58; water regulations in, 4–5, 154, 164–65, 202n4; water usage, 158, 159–62. *See also* legislative issues; *specific cities; specific states*
universal solvent, 25–29
unplanned reuse (de facto reuse), 112, 114, 115*fig.*
Unregulated Contaminants Monitoring Rule (UCMR), 66
unsafe water consumption, statistics on, 1–2
upgrading, economics of, 157
uranium, 51*tab.*, 172n4

value of exchange, 155–56
value of safe drinking water: meaning of, 155–59; public's role, 159–64
value of use, 155–56
Value of Water Campaign, 159–60
Van Vuuren, Lukas, 131
Venus, 6, 7, 8*tab.*
veterinary pharmaceuticals, 113*tab.*
Vibrio cholerae, 21–22, 21*tab.*, 35, 114*tab.*
Vietnam, 84, 149
violations, levels of, 157
viral diseases: eye infections, 21*tab.*, 114*tab.*; gastroenteritis, 21*tab.*, 114*tab.*; infectious hepatitis, 21*tab.*, 114*tab.*; respiratory illness, 21*tab.*, 114*tab.*; as waterborne disease, 21*tab.*
Virginia (USA), potable reuse and, 119*tab.*
viruses, 21*tab.*; Adenoviridae, 21*tab.*, 114*tab.*; Astroviridae, 21*tab.*, 114*tab.*; Caliciviridae, 114*tab.*; Hepeviridae, 21*tab.*, 114*tab.*

Washington, DC lead contamination, 74
wastewater: as contaminant source, 51*tab.*, 113*tab.*; discharge of, 50, 58–60, 112, 114; pathogens in, 114*tab.*
wastewater treatment systems: advanced water treatment, 121–25; conventional wastewater treatment, 120; CWA and, 50; economics of, 156; stress on,

wastewater treatment systems *(continued)* 161–62; from waste to potable reuse, 119–20. *See also* potable reuse

water: boiling point, 9*tab.*, 12–13; chemistry of, 3, 4, 164; heat capacity of, 15; melting point, 9*tab.*, 14–15; molecular formula of water, 128; paradox of, 3, 4, 164; solid water, 14–16; as universal solvent, 25–29. *See also* liquid water; molecular formula of water

Water Accountability and Equity Act (Baltimore, USA), 159

water availability, soil composition and, 102*tab.*

water bills, inequities and, 157

waterborne disease, 1–2, 4. *See also* cholera; typhoid fever

water conservation, as built infrastructure, 102*tab.*

water consumption: global water usage, 160*fig.*; in USA, 159–62

water contamination: about, 5; cancer risks from, 51*tab.*, 52; chemical contaminants, 81–87; chemistry and, 3, 155; factors effecting, 164–65; fertilizers and, 78–81; in Hoosick Falls (New York, USA), 2, 25, 67, 71, 130, 159; increase in, 3; in Newark (New Jersey, USA), 25, 73, 74, 77, 78, 159; precautionary principle and, 87, 155; processes of, 3; Roman Empire contamination event, 70, 72; understanding of, 4; US events of, 23, 71. *See also* chemical contaminants; Flint water crisis (Michigan, USA); lead contamination; nitrate contamination; PFAS contamination

water hardness, filtration and, 180n37

water improvement, economics of, 155

water infiltration, forest layers and, 102*tab.*

water management: chemistry and, 3, 5; economics of, 155; informed decisions for, 3; precautionary principle and, 5

water management services, comparison, 102*tab.*

water pricing structures, tiered pricing, 158

water quality: of aquatic systems, 161–62; education on, 162; global access and, 38–39; impacts on, 154; impacts on water quality, 32–35; importance of, 31–32; potable reuse and, 128; precipitation and, 3; socioeconomic benefits from, 42–44; sustainability and, 39–42; water security objectives, 102*tab.*; water storage strategies and, 30–31; water treatment benefits, 35–38. *See also* quality control

water rates: economics of, 157–58; tiered pricing, 158; understanding of, 163

water sector, careers in, 163–64

water security objectives, 102*tab.*; upgrading and, 157

watersheds: defined, 98, 99*fig.*; green infrastructures and, 106–9; indigenous communities and, 109; protecting of, 98–103; source contamination and, 97–98; watershed protection programs, 103–6. *See also* ecosystem health

water shortages: in New York City, 103–4; Value of Water Campaign, 159–60; water conservation, 160–62

water treatment: about, 4; chemistry and, 3; coagulation, 102*tab.*; economics of, 155; lead and, 179n27; microbial contamination and, 23, 154; point-of-use treatments, 137–48; SDWA and, 59*fig.*

water treatment plants: as built infrastructure, 102*tab.*; sediment removal, 102*tab.*

water usage: global water usage, 160*fig.*; in USA, 158, 159–62

water-use efficiency, as built infrastructure, 102*tab.*

water utilities: education on, 162; water rates, 163

waterway sediment, water security objectives and, 102*tab.*

WHO (World Health Organization). *See* World Health Organization (WHO)

Wilmington (North Carolina, USA), 71, 83*tab.*, 85, 97

Windhoek (Namibia, Africa), 119*tab.*, 126–27, 129, 131

World Health Organization (WHO): on arsenic levels, 146; on decentralization treatment, 134–36; drinking water standards, 125, 126; on global trends on cholera, 22*fig.*; on HWTS, 136, 146, 150; Joint Monitoring Programme for Water Supply, Sanitation and Hygiene (JMP), 40–42, 132, 135*fig.*; on lead levels, 54; on nitrate levels, 80; regulatory guidelines, 130; on safe drinking water, 1, 38

zinc oxides, 113*tab.*

Founded in 1893,
UNIVERSITY OF CALIFORNIA PRESS
publishes bold, progressive books and journals
on topics in the arts, humanities, social sciences,
and natural sciences—with a focus on social
justice issues—that inspire thought and action
among readers worldwide.

The UC PRESS FOUNDATION
raises funds to uphold the press's vital role
as an independent, nonprofit publisher, and
receives philanthropic support from a wide
range of individuals and institutions—and from
committed readers like you. To learn more, visit
ucpress.edu/supportus.